T0313034

Lighting & Controls:
Transitioning to the Future

Lighting & Controls:
Transitioning to the Future

Stan Walerczyk, CLEP, HCLP, LC

Routledge
Taylor & Francis Group

LONDON AND NEW YORK

Published 2020 by River Publishers

River Publishers

Alsbjergvej 10, 9260 Gistrup, Denmark

www.riverpublishers.com

Distributed exclusively by Routledge

4 Park Square, Milton Park, Abingdon, Oxon OX14 4RN

605 Third Avenue, New York, NY 10158

First published in paperback 2024

Library of Congress Cataloging-in-Publication Data

Walerczyk, Stan
  Lighting and controls : transitioning to the future / Stan Walerczyk, CLEP, HCLP, LC.
    pages cm
  Includes bibliographical references and index.
  ISBN-10: 0-88173-699-6 (alk. paper)
  ISBN-13: 978-1-4822-3683-5 (alk. paper)
  ISBN-13: 978-8-7702-2314-0 (electronic)
  1. Lighting. I. Title.

  TH7703.W29 2014
  621.32--dc23

                                                          2013041659

Lighting & Controls: Transitioning to the Future / by Stan Walerczyk
First published by Fairmont Press in 2014.

© 2014 River Publishers. All rights reserved. No part of this publication may be reproduced, stored in a retrieval systems, or transmitted in any form or by any means, mechanical, photocopying, recording or otherwise, without prior written permission of the publishers.

Routledge is an imprint of the Taylor & Francis Group, an informa business

Publisher's Note
The publisher has gone to great lengths to ensure the quality of this reprint but points out that some imperfections in the original copies may be apparent.

ISBN: 978-0-88173-699-1 (The Fairmont Press, Inc.)
ISBN: 978-8-7702-2314-0 (online)

While every effort is made to provide dependable information, the publisher, authors, and editors cannot be held responsible for any errors or omissions. The views expressed herein do not necessarily reflect those of the publisher.

ISBN: 978-1-4822-3683-5 (hbk)
ISBN: 978-87-7004-498-1 (pbk)
ISBN: 978-1-003-15183-8 (ebk)

DOI: 10.1201/9781003151838

Foreword

This lighting book should be a very useful hands-on tool for anyone involved with engineering, maintenance, purchasing, specification, retrofits, remodels, and new construction for interior or exterior applications. It shows very practical ways to maintain and update existing lighting systems and how to plan for the future. Key points include:

- LED is already cost effective in many applications and will become more prominent.

- Maintenance costs can be significantly reduced with fewer and longer life lamps and fewer lamp types.

- Very low power density can be achieved cost effectively while still providing very good lighting, such as with .4 watts per square foot in open offices, using ambient and task lights.

- The disadvantages of LED T8s, induction, dimming fluorescent ballasts, and reduced wattage T8s are given.

It is a challenge to balance past, existing, and future lighting. Some people want to buy LED products right away because of the WOW factor, even though incumbent technologies may be better. Others want to wait a year or two before doing a lighting retrofit, when LED products will likely have gotten better and less expensive. However, sometimes the lost savings by waiting are never recouped. This book will help people make good decisions.

Probably the most important part of the book is human centric lighting, also called human factors in lighting, biophilia, and other names. Human centric lighting (which includes the new and upcoming dimming and warm to cool white, color-changing LED products) may improve alertness, sleep, mood, visual acuity, energy savings, sustainability, and performance. Human centric lighting will probably be the next big step in lighting, perhaps as significant as Edison's creating the light bulb.

Table of Contents

Chapter 8

Induction . . . .

Chapter 9

Electronically Ballasted CMH . . . .

Chapter 10

Important Stuff To Know About LEDs . . . .

Chapter 11

Store Cooler Lights. . . . .

Acknowledgements

First, I want to thank my mother, who encouraged me and inspired creativity and fun in learning. She was also my best friend.

My father-in-law was a very good role model on how to do even unpleasant tasks gracefully. He was a longtime factory worker at Western Electric (which transformed into AT&T). He worked at the Hawthorne Plant, where the well-known Hawthorne studies were performed. He did not understand what I did for a living until he drove with me from the Chicago area to Des Moines, Iowa, and attended my full-day lighting class at the Iowa Energy Center. While driving back, he told me he understood why people paid to attend my classes. He died in early 2013 at 92 years of age, but he knew I was writing this book.

The most appreciation goes to my beautiful wife, who has graciously accepted my working 60-80 hours a week for decades and has made my life as simple as possible while writing this book. After finishing this book and easing into semi-retirement on Maui, I want to spend more time with her.

I also want to thank the lighting professionals who have helped me the most over the years. Without their help, this book would not be possible. The main two people in my early career were Brian Liebel, from whom I took every class he taught at the Pacific Energy Center in San Francisco, and Dr. Sam Berman, who developed spectrally enhanced lighting. As the years progressed, I felt fortunate to be able to work with them on the DOE spectrally enhanced lighting research and the IES Visual Effects of Lamp Spectral Distribution Committee.

I feel very fortunate to have had a cadre of lighting colleagues, including, in alphabetical order: Bill Brown, Brian Liebel, Brooks Sheifer, Dann Vail, Dallas Buchanan, Don Leaman, Don Link, Jeff Hirsch, John Fetters, John Hwang, Ken Patterson, Howard Wiig, Linda Hurd, Mark Whitney, Matt Tracy, Meredith Owens, Mike Lambert, Mike Smith, Mindy Haverland, Pat Becker, Robert Ofsevit,

Rod Heller, Sam Berman, Terry Clark, Tim Haley, Tom James, Tom Preston, and others. There are fond memories of two other lighting friends, Bart Wallace and Tom Tolen, who are no longer with us. Also, several of my clients have become good friends.

I also feel very fortunate to have had the opportunity to see so many intriguing facilities and to travel to so many wonderful destinations over the years for my lighting work. Such facilities include the back work areas of Disneyland, the San Diego Zoo, and Tesla Motors, the electric car company. Seeing various manufacturing processes has been a personal version of the TV show "How It's Made." Vacation-grade work destinations have included Alaska, Florida, Hawaii, Japan, New York City, and South Korea.

I would like to try to help others like the many who so graciously helped me.

Lastly, my 25th lighting anniversary was January 2, 2014, and I am so glad that this book was published during this significant year.

Chapter 1

Ready, Set, Go

This book is intended to provide engineers, owners, CEOs, facility managers, maintenance personnel, purchasers, contractors, ESCOs, and others with very practical information. Reading this chapter will help with understanding the book's perspective and purpose.

First of all, this is not a typical "engineering" book. It is more a first person commentary with solid facts, similar to the webinars and seminars that I present and the magazine articles and white papers I write. What do you expect from somebody who has a bachelors in psychology and a masters in oriental philosphy?

Although the AEE has wanted me to write a lighting book for years, I kept saying no because I thought a significant amount of information would be out of date within one or two years after publication. Some information will indeed be out of date by publication time. More has changed from 2010 to early 2014 than during the previous two decades, and the pace may quicken. But in 2012, I finally agreed, and I finished writing this book in early 2014 with the intention of providing sufficient tools and mindset to help readers keep up with the fast-paced evolution of lighting and make good decisions.

Although the AEE has wanted me to write a lighting book for years, I kept saying no because I thought a significant amount of information would be out of date within one or two years after publication. Some information will indeed be out of date by publication time. More has changed from 2010 to 2013 than during the previous two decades, and the pace may quicken. But in 2012, I finally agreed, and I wrote this book in 2013 with the intention to provide sufficient tools and mindset to help readers keep up with the fast-paced evolution of lighting and make good decisions.

While writing this book, I suspected that most people would be more prone to read particular chapters of interest for work, research, etc. than to read it from beginning to end. Thus, there is some duplication among chapters. While what is duplicated is important, and often seeing the same message more than once can be very beneficial, those who read this book one chapter after another can skip duplicated parts as desired. Much of the content consists of updated material from my various seminars, webinars, white papers, and magazine articles. For people who would like help with terms and definitions, there is both a listing of acronyms and a glossary in the appendix.

While I have consulted for various manufacturers, I do not accept any sales commissions. As a result, I have been called an "equal opportunity pisser-offer," because of a sense of fairness and a low tolerance for marketing hype.

Although many people push LED T8s, induction, dimming fluorescent ballasts, complex control systems, etc., this book will explain why they are not usually the best solutions.

There is a difference between features and benefits. Features include everything that a product can do, whereas benefits are what the customer really uses in a productive manner. For example, Microsoft Excel has an incredible array of features, but most people use only a small percentage of that capacity. So why buy a very complex control system if there are idiot-proof and localized occupancy sensors to do the job? One rule of thumb I have found very useful over 24 years in lighting is "follow the money." (Is somebody recommending something to you because it is best for you, or because he or she will make the most money from it?)

This is not one of those lighting books written by an "ivory tower expert" who never got on a ladder to replace lamps and ballasts; never dealt with late, damaged or wrong shipments of products; or never had to make a sale of a lighting retrofit project. Nor is it written by an Orwellian "big brother" who mandates unrealistic requirements for rebates, building codes, etc. Such people often think they know more than lighting professionals and end customers on a specific project. For those who are interested, my bio and contact information are in the appendix. (Although some people consider me opinionated, I can back up my positions!)

It will become an LED or other solid state lighting (SSL) world, especially with over 100 LPW out of fixtures and 10-year warranties. The major challenge is to balance the maintenance of existing lighting and controls with the use of currently available high performance lighting—including fluorescent, electronically ballasted CMH, LED, and wireless controls—to build or retrofit, while projecting what will be available in the future. Making choices about lighting is even more difficult when funds are very limited. Various manufacturers, agents, distributors, lighting retrofit companies, ESCOs, and others may give you opposing recommendations. This book should help you make sense of where they are coming from and give you tools for developing your cost effective strategies.

One of the big choices in a retrofit, remodel, or new construction is going with high performance fluorescent T8 systems, or LED troffer kits or troffers. The pros and cons of each will be discussed herein. (Please avoid LED T8s.)

I read Thomas Friedman's and Michael Mandelbaum's book *That Used To Be Us—How America Fell Behind in the World It Invented and How We Can Come Back*. One of the major messages is that average does not cut it anymore. I keep running into lighting retrofit contractors, ESCOs, engineers, lighting designers, and end customers who want to keep doing lighting retrofits, remodels, and new construction design the same they have for years "because it has worked well."

But this is a new lighting world, and if people keep old habits, they may not keep their jobs. (On the other hand, it is not always good to jump into new technologies. An example is LED T8s. Early adopters are needed to really learn how new products work, but often tried and true solutions are safest for many people and organizations.)

Reading this book should make it easier to understand more depth and width from the constantly upcoming webinars and seminars that I and others present for AEE and other organizations. Some website addresses are listed, but as you probably know all too well, these addresses can change over time, so you may have to do searches with key words.

Probably the most important part of the book is human centric lighting, also called human factors in lighting, biophilia, and other names. Human centric lighting (which includes the new and upcoming dimming and warm to cool white, color-changing LED products) may improve alertness, sleep, mood, visual acuity, energy savings, sustainability, and performance. Human centric lighting will probably be the next big step in lighting, perhaps as significant as Edison's creating the light bulb. If you only read one chapter, please read Chapter 29 on human centric lighting; it really shows that lighting is much more than a commodity.

I hope you like the rest of this book. If there is sufficient demand, it could be revised in the future.

Chapter 2

Yesterday, Today, And Tomorrow

There are not fixed lines among the lighting of yesterday, today, and tomorrow, because they often overlap. Today, you may have some of all three types of lighting.

Yesterday's lighting is not only T12s with magnetic ballasts but also what is called first generation, basic grade, or 700 series full wattage T8s with generic electronic ballasts, which are still very common. The USA will probably stop production of basic grade T8s in the summer of 2014, and Canada and other countries may follow. Production of generic electronic ballasts may sometime also be eliminated. CFLs, even with electronic ballasts, and HID lamps with magnetic ballasts are also yesterday's lighting. If the KWH rate is at least a nickel, it is usually cost effective to replace these lamps and ballasts. But there may not be sufficient money or time to do it soon. Some people replace basic grade full wattage F32T8s with 28W or 25W F32T8s, lamp for lamp, while keeping the existing ballasts. While this does save some wattage, it often makes later optimal solutions not cost effective.

Today's lighting includes high performance T8s with high performance ballasts, and electronically ballasted CMH and LED, some of which are tunable. Although numerous people promote dimming fluorescent ballasts, these are not energy efficient because dimming below .70 BF requires heating lamp cathodes, which consumes significant wattage. Also, although numerous people are pushing controls, they are often not cost effective because the power density and electric costs are so low with high performance lighting.

Tomorrow's lighting will be LED or another type of SSL, mainly tunable with integral wireless controls, which are considerably less expensive than existing controls. Today, some people want to wait until LED products and controls are improved and less expensive before doing a retrofit or upgrade, but often the lost savings from waiting are never recouped. It is tricky making the best decisions, but knowing what type of lighting you currently have can be helpful.

Since I started writing this book in December of 2012 through early 2014, today has actually started the shift to tomorrow, with LED becoming really cost effective for troffers, task lights, recessed cans, accent lighting, decorative lights, refrigerators and freezers in grocery stores, hibays, garages, exterior lights, etc.

Chapter 3

Lighting 101

INTRODUCTION

This is a very short chapter. For those readers needing to learn lighting basics, I highly recommend either taking my Lighting 101A, B, C & D 24/7 online seminars through AEE, or another basic lighting class from another person or organization. The glossary and list of acronyms in the appendix may also help.

BASIC MESSAGE

Lighting can often be the easiest and most cost effective way to save energy in residential and commercial facilities. It is often the "cash cow" to subsidize other energy measures, such as new HVAC or motor controls. Sometimes savings from lighting can help pay for new roofs and other needs. Good lighting can also improve worker satisfaction, worker productivity, student performance, and retail sales. Perhaps you like or dislike some lighting but do not know exactly why or how to communicate with others about it. Reading this book should help.

What is light? It is a very narrow band from about 380 to 770 nanometers in the electromagnetic scale, bordered by ultraviolet on one side and infrared on the other. (See Figure 3-1.) Certain lighting technologies, especially incandescent, have a lot of energy in the infrared area, which is why such lamps create so much heat. In fact, incandescents are better heaters than lighters. Before 100W incandescent bulbs were eliminated, two of them were used in Easy Bake Ovens.

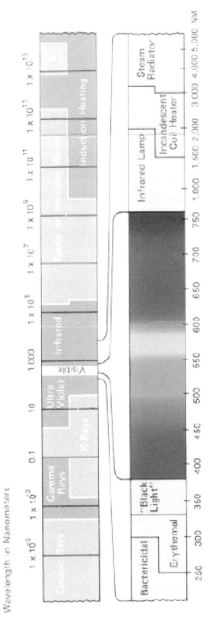

Figure 3-1. Electromagnetic spectrum

Chapter 4

Maintenance

INTRODUCTION

In my opinion, if many of the original lighting designers and architects visited the site after an operating period of at least four years, they might not make it out without having things thrown at them. I also think that if lighting designers and architects had to do maintenance, they would not:

- Include a zillion lamp types, some of which may be expensive, single-source items and not be the longest-life lamps

- Use way too many lamps

- Specify fixtures that present a challenge in getting to lamps and ballasts

- Have fixtures installed in hard to reach locations, such as over stairs

Especially since China started to jack up the price of rare earth materials in the summer of 2011, and those materials are used in phosphors in fluorescent lamps, using fewer lamps saves money on purchasing, recycling, and labor. For example, most 2x4 troffers should have one or two lamps, not three or four.

LAMPS TO BE ELIMINATED OR AVOIDED

In general, the following are lamps that should be eliminated or avoided because of high cost, low LPW, bulkiness, being often not locally stocked, and/or short life:

- Most incandescents and standard halogens
- T12s
- 3', 5', 6', and 8' lamps
- U-bend lamps
- Biax lamps
- T5HOs
- Most pin based CFLs
- Mercury vapor lamps
- Standard or probe start MH lamps
- Induction lamps

It has been my experience that maintenance people do not like bulky lamps such as eight footers and U-bends. One facility wanted to get rid of U-bend T8 lamps, mainly because their personnel could not fit very many of the lamps on their carts with all of the other lamp types. Plus, the lamps were expensive to buy and recycle and did not last that long. Fixtures with two U-bend T8s or T12s can usually be retrofitted with 2' F17T8s.

Also based on my experience, most maintenance people hate 8' T8 and T12 lamps. These lamps are difficult to store and transport. If you have ever been on a ladder, lift, or scaffold, you know that dealing with 4' lamps is a lot easier than with 8' lamps. Most 8' fixtures can easily be retrofitted or replaced with end-to-end 4' lamps. For example, an 8' strip fixture with two 60W 8' F96T12 lamps and magnetic ballast can be retrofitted with an 8' strip or hooded industrial kit, two high performance 32W F32T8s, and high performance .87-1.20 BF electronic ballast. Also, two 4' F32T8s cost less than one 8' T8 or T12 lamp.

Since the volume is so low, many distributors do not even stock 3' and 5' lamps, and they are usually expensive. For example, a 3' T8 lamp costs significantly more than an equivalent 4' T8. The T5 lamps and ballasts are relatively good but not usually as good as high performance T8 lamps and ballasts. T5 lamps and ballasts also cost significantly more. Also, most T5 and T5HO lamps are manufactured in China or other countries, while most T8 lamps for North America are manufactured in the United States and Canada, which of course is better for jobs in these countries.

Without reducing performance, single-source lamps should be avoided because suppliers often charge more for them.

WORKHORSE LAMPS

A list of workhorse incumbent technology lamps includes:

- High performance 4' 32W F32 T8s
  — highest lumen and long life
  — extra long life and mid lumen

- High performance 2' F17T8s where 4' lamps will not fit

- If pin based CFLs will be used, try for a maximum of two types in a building

- Pulse start, preferably ceramic MH lamps with electronic ballasts

In regard to highest lumen and long life, 32W F32T8s are on the CEE's high performance list. There may only be one extra long life and mid lumen 32W F32T8 that CEE approves, which is the 5000K version of the Sylvania XP XL. Sometimes it is good to have extra long life T8s, even if they do not qualify for the CEE list and rebates, because they can last up to 67,000 hours at 12-hour cycles with program start ballasts, which can really reduce maintenance costs.

Although some lamp manufacturers, retrofit contractors, ESCOs, and rebate organizations have been pushing reduced wattage 25W and 28W F32T8s, these lamps are really not that good. They usually should not be used below 60°F in hibays, with many older generic electronic ballasts, nor with many older dimming electronic ballasts. Since they have less lumens, they are not nearly as good as high performance 32W F32T8s with delamping.

Also, the reduced wattage T8s have a lower LPW than the highest lumen 32W versions, when equivalent ballasts are used.

HPS lamps may be okay for a while for exterior applications, until LED and electronically ballasted CMH gets better and less expensive. Although most HPS lamps are rated for 24,000 hours, some are rated for 30,000-40,000 hours.

LAMP LIFE

For practically everything other than LED, lamp life is based on when half of the lamps are still working and half of the lamps have burned out in laboratory conditions. This can easily be seen in a bell shaped curve, with the top of the bell being the rated life. (See Figure 4-1.) A few lamps may burn out in a few months, and some may last over a decade, but most will burn out at 70-130% of rated life.

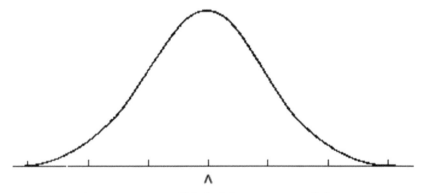

Figure 4-1. Lamp life in laboratory conditions

Mercury vapor lamps are primarily rated for 24,000 hours, but they usually last considerably longer, getting considerably dimmer over time. Similarly, 20,000-hour rated, standard 400W MH lamps last longer and also get dimmer. If the manufacturers listed longer-rated lives, the mean lumens would be worse than they are now, which is already not very good.

RELAMPING

Spot Relamping

Most people and organizations relamp when lamps burn out, but it is usually quite expensive. A number of years ago, I visited the facility manager of a hospital and asked him how much it costs the hospital to do a spot lamp replacement. He did not know but was open to finding out. He knew that there was at least one bad lamp in a certain room on the third floor. Noting the time of day, he called the person responsible for replacing lamps during that shift and told him to change the lamp right then and to let him know when he had finished. The person in the maintenance area in the basement went up to the third floor, got a ladder, saw the fixture with the burned out lamp, and opened the fixture to get the exact lamp type. Then he went back to the basement to get a new lamp, brought it back up to the third floor, replaced the lamp, put the ladder back in the third-floor closet, and brought the burned out lamp back to the basement. This took 20 minutes. Based on $45/hour, including benefits, that was a $15 labor cost for that one lamp.

Group Relamping

Group relamping is usually much more cost effective than spot relamping. The labor for group relamping may only cost $2 to $3 per lamp. Also, with the higher volume, unit lamp pricing and lamp recycling costs are usually lower. Often parts and labor, including cleaning and recycling, may cost about $5 per lamp.

At 70-80% of rated lamp life is usually the best time for group relamping because, while not that many lamps have been burned out, a substantial number of them will burn out soon if not replaced. The best way to schedule a group relamping is to determine the hours of use after the original lamps were installed or after they were replaced in the last retrofit. For example, if the lamps have a 20,000-hour rated life, 70-80% of that is 14,000-16,000 hours. Based on most of those lamps lasting 3500 hours a year, the group relamping should be scheduled four to four and a half years out. Funding for this should often start about two years ahead of time.

In buildings that are relatively old and in which no retrofit has been recently done, an average of half-life can be considered for existing lamps. For example, 20,000-hour rated lamps can be considered to have an average of 10,000 hours remaining; with that information, a group relamping or a retrofit can often be cost effective.

Relamping HID Hibays

It is important to know how a facility replaces its MH, HPS, or mercury vapor lamps in hibays. Do they use a pole changer from the ground or from something higher, like from the top of a forklift? Do they have their own lift or scaffold? Do they rent a scaffold? Do they hire a company? Costs can vary widely, depending on these factors.

LONG LIFE LAMPS

With either spot or group relamping, long life lamps are usually recommended. There are long life incandescent, halogen, and halogen infrared lamps, which are usually 130V versions used at 120V. But be aware that LPW gets worse with those lamps. Although most HPS lamps are rated for 24,000 hours, there are good 30,000-40,000-hour rated options. There are numerous long life MH lamps rated for 30,000 hours and longer.

In the fluorescent world, lumens and life fight against each other. Up to a certain point, with better materials, both lumens and life can be improved. But at a certain level, if maximum lumens are desired, life must be sacrificed, and if maximum life is desired, lumens must be sacrificed. So, you can either have highest lumens or longest life. In Table 4-1, the 32W GE HL, Philips ADV, and Sylvania XPS lamps are the high lumen lamps, while the 32W GE SXL, Philips ADV XLL, and Sylvania XP/XL are the

Table 4-1

4' T8 LAMP LIFE, LUMENS, CRI & MERCURY

LAMP	WATTS	3000-4100K		5000K		MAX MG OF HG	LAMP LIFE HOURS			
							INSTANT START		PROGRAM START	
		LUMENS	CRI	LUMENS	CRI		3 HR	12 HR	3 HR	12 HR
1st GENERATION - GENERIC	32	2800	75-78	2800	75-78	1.7 - <10	15,000 - 24,000	20,000 - 30,000	20,000 - 30,000	24,000 - 36,000
2nd GENERATION - GENERIC	32	2950	81-85	2800 - 2950	80-85	1.7 - <10	15,000 - 24,000	20,000 - 30,000	20,000 - 30,000	24,000 - 36,000
GE HL	32	3100	82	3000	80	2.95	25,000	36,000	40,000	45,000
GE SXL	32	2850	81+	2750	80	2.95	31,000	40,000	55,000	60,000
PHILIPS ADV	32	3100	85	3000	82	1.7	24,000	30,000	30,000	36,000
PHILIPS PLUS	32	2950	85	2850	82	1.7	30,000	36,000	38,000	44,000
PHILIPS ADV XLL	32	2950	85	2850	85	1.7	40,000	46,000	46,000	52,000
SYLVANIA XP	32	3000	85	3000	85	3.5	24,000	40,000	40,000	42,000
SYLVANIA XPS	32	3100	85	3100	81	2.9	24,000	40,000	40,000	42,000
SYLVANIA XP/XL	32	2950	85	2950	81	3.5	36,000	52,000	65,000	67,000
GE SPX 28W	28	2725	82	2625	80	2.95	24,000	30,000	45,000	50,000
PHILIPS ADV 28W	28	2725	85	2675	82	1.7	32,000	38,000	38,000	44,000
PHILIPS ADV XLL 28W	28	2675	85	2625	82	1.7	40,000	46,000	46,000	52,000
SYLVANIA XP 28W SS	28	2725	85	2725	81	2.9	24,000	40,000	40,000	42,000
SYLVANIA XP XL 28W SS	28	2600	85	2600	81	5	50,000	75,000	80,000	84,000
GE SPX 25W	25	2400	85	2350	80	2.95	36,000	40,000	50,000	55,000
PHILIPS ADV 25W	25	2500	85	2400	85	1.7	32,000	38,000	38,000	44,000
PHILIPS ADV XLL 25W	25	2400	85	2350	82	1.7	40,000	46,000	46,000	52,000
SYLVANIA XP 25W SS	25	2500	85	2500	81	2.9	24,000	40,000	40,000	42,000
SYLVANIA XP XL 25W SS	25	2400	85	N/A	N/A	5	50,000	75,000	80,000	84,000
F28T5	25-28	2900+	85	2750+	85	1.4 - 2.5	*	*	25,000 - 30,000	30,000 - 40,000
F54T5HO	49-54	5000	85	4800+	85	1.4 - 2.5	*	*	25,000 - 30,000	30,000 - 60,000

Lamp manufacturers may alter rated lamp life and lumen specifications, so get updates from manufacturers.

Prepared by Stan Walercyk of Lighing Wizards   1/1/14   version   www.lightingwizards.com

extra long life lamps. You can see how they compare. As stated in the table, periodically check with manufacturers, as specifications can change.

REBALLASTING

Most magnetic and electronic ballasts are rated for about 60,000 hours; just like with lamps, the middle of a bell shape curve represents when half of the ballasts are burned out and half are still working. That is often about 15 years. Often, ballasts start to burn out in significant numbers in about 10-12 years.

Magnetic ballasts should definitely be eliminated because they are inefficient. Even generic electronic ballasts, the main electronic ballasts over five years ago, should be replaced with equivalent high performance electronic ballasts, saving 3-6 W. Each high performance electronic ballast can save $15-$40 in electricity (depending on KWH rate) over the rated life compared to a generic electronic one, and even more compared to a magnetic one.

Ballasts that are 10 or more years old are prime candidates for retrofit projects. Usually, if there is no retrofit when they burn out, they are simply replaced with similar ones on a spot basis, which is expensive regarding labor and the usual pricing on a small number of ballasts. With a retrofit, there will be energy savings, labor savings, unit parts savings, and all new ballasts with a five-year parts and labor warranty.

Also with a retrofit, especially with delamping, the wattage is significantly reduced, which directly translates to less heat, the enemy of all electronics, including electronic ballasts. The best way to evaluate ballast heat is the case temperature, the hottest part of the outside of the ballast. It can get quite hot in a ballast compartment from the heat of the ballast and lamps, especially in an enclosed fixture. At 70°C (158°F) ballast case temperature, the rated ballast life is often listed at 60,000 hours. At about 10°C (18°F) higher ballast case temperature, ballast life may be reduced to 30,000 hours, while at about 10°C (18°F) lower ballast case temperature, ballast life may be increased to 90,000 hours. For example, delamping a 4-lamp fixture with either two 2-lamp ballasts or one 4-lamp generic electronic ballast down to two high lumen lamps with one 2-lamp high performance ballast should significantly reduce temperature and increase ballast life.

There are some fluorescent lamps rated for 84,000 hours and longer, but since most ballasts may die before that, is that long lamp life really useful? By the time new ballasts in recent new construction (or those in a retrofit) really start to burn out, we will be in an LED or other type of SSL world.

CLEANING

The need for cleaning really depends on how dirty the lamps, reflective surfaces, and one or both sides of the lenses get, as well as the type of dirt involved.

Offices, Classrooms, etc.

Let's consider offices, schools, stores, etc. first. Now that smoking is generally banned in these areas, and there is often filtered air, cleaning is not nearly as necessary as in the past. Often it is okay to only clean the fixtures when lamps are replaced. But now, with extra long life fluorescent lamps, which last up to 67,000+ hours with program start ballasts and 12-hour cycles, it may be good to clean fixtures once in between replacements. (It is easy to measure footcandles before and after cleaning to see how much of a difference cleaning makes.)

Quite a while ago, a Silicon Valley company called me because office workers were complaining about too little light. The company had suspended, totally indirect lighting fixtures. After I climbed up a ladder, the problem was very obvious. There was so much dirt on the bottom specular reflector that I could not even see the reflector. When I asked when was the last time the fixtures were cleaned, I was told never—and the building had been built about 10 years before! After wiping off the thick dirt dust, the light levels were fine. I instructed the company to clean all of those fixtures throughout the building immediately, and that down the road, the fixtures should also be cleaned when lamps are replaced. In general, I prefer suspended fixtures with some type of open bottom so dirt does not accumulate that badly.

Hibays

Many industrial and other facilities can be much more of a problem, with lighting fixtures requiring more frequent and more proper cleaning. Again, it is important to know what kind of dirt. There is dirt that settles on top of stuff, there is static dirt that can be attracted to any surface, and there is grimy dirt that can stick to all sides of objects.

A number of years ago, when I worked for a lighting retrofit company, the ESCO specified that we install MH retrofit kits with an inner reflector, in addition to the larger outer one, in hibays in a large printing shop. The lighting was quite good initially, but especially problematic was the relatively small inner reflector, on which this system heavily relied. It got so oily and grimy from the ink in the air that light levels got noticeably lower within a month, and cleaning was needed almost every month. A lighting system that relied less on reflectors would have been better, even if more wattage was required.

Mainly for hibays, but also for some other fixtures, glass and acrylic domes are two options. Yes, glass costs more and weighs more, but it does not get static charges, so it can stay cleaner.

LED IS DIFFERENT

With the advent of LEDs, the entire world will have to change the way maintenance is done. LEDs don't really burn out; they just get dimmer and dimmer over time, like mercury vapor.

I have been in numerous industrial buildings and warehouses with 1000W mercury vapor hibays, which only provide about five footcandles. I have often told the maintenance people that this means their site is underlit for work productivity and safety, but they point up to the hibays and state that the lamps are on, so they will not do anything.

There are millions and millions of LED exit signs and kits that are more than ten years old. Many of these no longer provide sufficient light for National Fire Protection Agency (NFPA) or other codes. Again, I point this out to maintenance people, and they usually tell me the signs are working, so they are not going to do anything.

Hopefully, it will not take something like people dying in an industrial accident or from not getting out of a burning office building fast enough due to too little light, scenarios which could also result in massive lawsuits and maybe even destroy some companies and organizations. Here is one good way to prevent this:

1. Check required light levels in critical areas per the IES, NFPA, OSHA, etc.

2. Research, specify, purchase, and install LED fixtures, which provide at least 30% higher lighting than those required light levels.

3. Check light levels about four years after installation, or sooner if lighting seems dim.

4. Then start checking light levels every year or two.

5. When light levels are only about 10-15% above required light levels, note the date and start efforts to get money to retrofit or replace all the fixtures. (It usually takes at least two years to get enough money allocated.)

6. When light levels are close to the minimum allowed, it's time for the retrofit or replacement to be done.

Maintenance is also an important concern in non-critical areas such as offices and classrooms.

Good fluorescent lamps only lose about 10% of initial lumens at end of rated life. Then they burnout and are replaced, and hopefully the fixtures are cleaned as well, bringing light levels back up to 100%. But the life of most interior LED troffers, troffer kits, and other projects are based on L70, which means losing 30% of initial lumens at the end of rated life. Thus, if there is no control system, the area will either be overlit to begin with and have proper lighting at the end of rated life, or the area will have proper lighting initially and be underlit at end of rated life. Neither is good.

There are some LED solutions as good as or even better than the best fluorescent ones regarding lumen maintenance. Finelite HPR LED troffers only lose 10% of initial lumens at a rated life of 100,000 hours. Acuity Brands' nLight system under-drives the drivers about the first half of rated life and then increases drive current as the LEDs become less bright. Photocontrol systems could be used. Digital clock chips could be included to slowly increase drive current over time.

In general, L70 is okay for exterior LED fixtures, because HPS and MH can lose that much light. However, it would be good to go to L90 (rated life being when 10% of initial lumens are lost) for interior LED fixtures.

Chapter 5

Dimming Fluorescent Ballasts

This very short chapter's message is: Do not use dimming fluorescent ballasts for attempted energy savings because they are:

- Expensive.

- Energy hogs, especially when dimmed below .70 BF when they have to heat lamp cathodes so that lamps do not flicker, spiral, or turn off.

- Mainly series wired, so if one lamp burns out, it will take at least another one out with it.

Dimming every dimming fluorescent ballast to 50% consumes about 20-30% more wattage than turning off every other equivalent fixed output ballast. Such low power densities can be achieved with high performance fixed output fluorescent and LED systems, so even if fluorescent dimming ballasts save some energy, they and their controls are usually not cost effective. The future of dimming will be with LEDs because they can maintain or get more efficient when dimmed, since they run cooler then.

Chapter 6

Spectrally Enhanced Lighting

INTRODUCTION

Spectrally enhanced lighting (SEL) is one of the best and most underutilized techniques to save energy while maintaining or improving visual acuity and preserving wattage. Most of the time it is used to do merely the former.

BACKGROUND

Dr. Sam Berman developed scotopically enhanced lighting (which became spectrally enhanced lighting) after intrinsically photosensitive retinal ganglion cells (ipRGCs) were discovered in 2002. Brian Liebel managed three phases of research on this for the DOE. Sam, others, and I assisted. Some of the case studies are mine. For information, see:
http://www1.eere.energy.gov/femp/technologies/eut_spectral_lighting.html

Brian is the chair of the IES Visual Effects of Lamp Spectral Distribution Committee. Sam, numerous others, and I have been on this committee for at least four years, assisting with the writing of Technical Memorandum 24 (TM-24).

About 90% of my retrofits over the last 15 years have been with 5000K or higher CCT fluorescent, and that should continue with LED. Numerous lighting retrofitter contractors and ESCOs have been using 5000K fluorescent for years, and they will probably also transition to 5000K LED.

TM-24-13

During the first half of 2013, the IES approved TM-24-13 with one of the longest titles I have seen: "An Optional Method for Adjusting the Recommended Illuminance for Visual Demanding Tasks Within IES Illuminance Categories P through Y Based on Light Source Spectrum." It is highly recommended that lighting people read this document. The cost may be $50 to non-IES members. (See: www.iesna.org) So now, finally, the IES approves of the benefits of spectrally enhanced lighting in demanding tasks.

ENERGY SAVING OPPORTUNITIES

The reference light source is 3500K, which is considered to have a scotopic/photopic (S/P) ratio of 1.40.

For demanding tasks that are not overlit, the following are lower photopic light level percentages with higher CCT lighting while maintaining visual acuity, compared to 3500K.

- 14% with 1.70 S/P ratio 4100K
- 23% with 1.95 S/P ratio 5000K
- 31% with 2.22 S/P ratio 6500K
- 36% with 2.45 S/P ratio 8000K

Since wattage can be a linear one for one relationship with light levels, the above are also methods of potential energy savings. However, be aware that 6500 and 8000K lamps have reduced photopic or catalog lumens.

With a fluorescent T8 system, it is often good to break it down into BFs and delamping. Some examples with equivalent photopic lumen lamps and typical fixed output ballasts are:

Lamp for Lamp
- Two 3500K 32W F32T8s & 1.15 BF high performance IS ballast consumes about 73W.

- Two 4100K 32W F32T8s & 1.00 BF high performance IS ballast consumes about 64W.

- Two 5000K 32W F32T8s & 0.87 BF high performance IS ballast consumes about 54W.

Delamping Opportunity
- Two 3500K 32W F32T8s & 0.77 BF high performance IS ballast consume about 48W.

- One 5000K 32W F32T8 & 1.15 BF high performance IS ballast consumes about 38W.

Although LEDs are also available in 3500, 4100, 5000 and higher Kelvins, the S/P ratios could be quite different for the same CCT fluorescent lamps and LEDs. Even LEDs of the same CCT may have considerably different S/P ratios. Hopefully, the DOE will test LEDs in 2014, like it started testing fluorescent in 2013.

EVEN THE BEST CHANGE SINCE SLICED BREAD

Some people do not initially like the best change since sliced bread, no matter what it is, because they are used to the status quo; it can take a while them for make the internal adjustment with a change. We really saw this in the first phase of the DOE research. A number of people who were used to 3500K did not like the change to 5000K, and they wrote that in the initial questionnaire.

But, after getting used to the 5000K after three weeks, many of those people became the biggest supporters of the new lighting, as shown in that questionnaire. After three weeks, one's comment was basically, "I am glad I have the 5000K. When I go to another floor that still has the 3500K lamps, it looks dingy and yellowish." It turned out that for all office workers, after three weeks there was no statistical preference of 3500K over 5000K.

Some people have stated that this is due to the Hawthorne Effect, but that is inaccurate, based on the scientific methodology. (As many readers already know, The Hawthorne Effect refers to a situation where subjects modify their behavior because they know they are being studied. The term comes from the Western Electric Hawthorne factory outside Chicago, where productivity improved when lighting was made brighter or dimmer, then slumped when the study was over. In fact, my father-in-law worked there, but that was after the study was done.)

Having learned from "the school of hard knocks," I can now avoid the problems that used to happen. Before a retrofit project, I have an email sent out to all workers, stating:

- There will be a lighting upgrade. (Sounds much better than "retrofit.")

- The new color tone will be more like daylight. (Who does not like daylight?)

- The new color tone is based on spectrally enhanced lighting, which is endorsed by the DOE as an excellent way to reduce energy, carbon footprint, etc.
  — Provide weblinks to the DOE.

- If anybody does not like the new lighting initially, please keep it to yourself.

- If anybody does not like the new lighting after two or three weeks, please let me know.

It is also important for the installers to finish rooms before the regular workers come for the next day's work because having two color tones in one space does not look good. With this protocol, I have not had to change the lighting in even one office back to a warmer CCT.

Many lighting designers do not like high CCT fluorescent lamps and usually specify 3000K or 3500K fluorescent in new construction and remodels. I do not understand this, because most of the light designers like daylight, which is considered an average of 6500K. Maybe it is because cool white fluorescent has considerably less red content than warm white fluorescent, which does not have that much red anyway.

Most retrofits done with at least 4100K (and often 5000K) have good customer satisfaction. High CCT LEDs could be developed with substantial R9, which is red content, and that may help many more people like, or at least accept, high CCT. As seen in the Human Centric Lighting chapter (29), high CCT lamps also have additional benefits.

Chapter 7

Delamp vs. Lamp-for-lamp Retrofit

INTRODUCTION

This chapter is about going from T12 and basic grade T8s to high performance fluorescent T8 retrofits.

For the numerous lighting retrofit contractors, ESCOs, distributors, consultants, end customers, rebate administrators, and others that have been mainly promoting and specifying lamp-for-lamp retrofits with 28W or 25W F32T8s, or minor delamping with those lamps, please consider the advantages of major delamping, especially with high lumen, which is 3000-3100 initial catalog or photopic lumen, 32W F32T8s. This is especially important since lamp prices have significantly increased since the summer of 2011, when China jacked up prices on rare earth materials used in fluorescent phosphors.

Although 28W and 25W F32T8s can be used for limited delamping, high lumen 32W F32T8s usually provide the lowest number of lamps, highest energy savings, and lowest maintenance costs. Many fixtures with up to three basic grade T8s can be delamped down to one high lumen 32W F32T8 with upscale kits. There are one-lamp reflector kits for two-lamp 2x4 lensed troffers; two-lamp kits for fixtures that have two 60W F96T12s, and numerous other kit and fixture types.

Kits are necessary for delamping many fixture types; however, numerous fixture types, such as four-lamp 2x4 lensed troffers with interior angled long sides and good cleanable white reflective surfaces, as well as two-lamp 4' strip fixtures, that can be delamped with or without kits. There are pros and cons either way.

The high lumen 32W F32T8 is the best workhorse fluorescent lamp, even in non-delamping applications.

COMMON REFLECTOR SYSTEMS

Consider two samples from Energy Solutions International. Figure 7-1 shows how a three- or four-lamp troffer was retrofitted with a two-cove reflector and two lamps. Figure 7-2 shows how a two-lamp troffer was retrofitted with a one-cove reflector and one lamp.

Figure 7-1. Retrofit with a two-cove reflector

Figure 7-2. Retrofit with a one-cove reflector

REASONS FOR LAMP-FOR-LAMP RETROFITS

During my seminars, consultations, etc., I have been told various reasons to go lamp-for-lamp with 28W or 25W F32T8s. Following is some of what I have heard:

- We look at auditing as a necessary evil (necessary to keep the installers working), and we can audit this way the fastest.

- A qualified lighting professional is not required for the audit and specification.

- This way we do not have to measure, order, receive, and install any kits, so we can work faster with fewer potential headaches.

- As a lamp supplier, our sales volume stays high and we sell these lamps for spot or group relamping.

- A retrofit contractor or an ESCO is not always necessary.

- The best rebates are with these lamps.

- The new lighting is very similar to the old lighting, so nobody should complain.

Regarding the last bullet, if the old lighting was crappy, the new lighting can still be crappy. In general, all of these reasons are not very good.

LAMP COMPARISONS

It is my opinion that reduced wattage T8s are better for the lamp manufacturers and distributors than for end customers because the profit margins are good and sales volume is kept high.

Many people think they can save 4W with a 28W F32T8 and 7W with a 25W F32T8 compared to a basic grade 32W F32T8, but the savings when driven by most high performance electronic ballasts are more like 2.5 with 28 watters and 5 with 25 watters.

Also, 28W and 25W F32T8s should not be used below 60°F and do not work well with many older fixed output and dimming electronic ballasts.

One advantage with 28W and 25W F32T8s in certain hot applications is that they have optimal light output, at 92-95°F, compared to 77°F for 32W F32T8s. But with delamping, which usually results in higher fixture efficiency and lower wattage (translating to less heat), the high performance 32W F32T8 systems can usually outperform reduced wattage T8 systems.

An interesting note is that four 25W F32T8s with high performance .77 ballast factor (BF) ballast consume about the same wattage and provide about the same simple-calculated lumens as two high lumen 32W F32T8s with high performance 1.15-1.20 BF ballast. Including fixture efficiency with fewer lamps, the two full wattage lamps and high BF ballast system usually provides more light. Who wants to have to buy, install and recycle two extra lamps down the road? Also, high lumen 32W F32T8s and similar life 28W F32T8s cost about the same, while similar life 25W F32 T8s cost more. Plus, reduced wattage T8s usually do not have sufficient lumens to be used in hibays and other fixtures.

High lumen 32W F32T8s have about the same or better lumens per watt than 28W or 25W F32T8s, when used with the same electronic ballasts. High performance 32W F32T8 systems also have higher lumens per watt than T5 and T5HO systems. Just dividing initial lumens by lamp wattages from lamp catalogs gives T5s, T5HOs, and reduced wattage T8s an artificial advantage over high lumen 32W F32T8s because the 32W F32T8s are tested with reference magnetic ballasts, while the newer lamp types are tested with reference electronic ballasts that are more efficient. The 28W F32T8s with 1.00 BF ballast provide about as many lumens at about the same wattage as high lumen 32W F32T8 with .87 BF ballast, but 1.00 BF ballasts can be more expensive.

For T12, T8, and T5 fluorescents going from standard output to high output there is an increase in lumens but a reduction in lumens per watt. High performance T8 systems are 8-18% more efficient than T5HO systems. In the fluorescent world, lumens and life fight against each other. So, you can get the highest lumens or the longest life, but not both. Sometimes saving every watt is the most important; sometimes saving the most on maintenance is the most important. After China really started jacking up the price of rare earth materials necessary in phosphors, at least one major lamp manufacturer has introduced reduced wattage T8s with less phosphor in order to keep the pricing down, but these lamps have less lumens.

Table 7-1 should be helpful. Columns B-D represent information that can be attained from a lamp catalog, but the numbers given are not that good. Columns E-N are much more realistic. I used two-lamp systems, but one-, three-, or four-lamp systems could be used. The most important columns are M and N. The lower phosphor lamps are not included.

WORKHORSE LAMPS

High lumen 32W F32T8s are the best overall linear fluorescent lamps, not only because of efficacy. Advantages include:

- Good when lamp-for-lamp can use down to .60 BF fixed output ballasts

- Can be used with all older, fixed output and dimming T8 ballasts

- Usually better than T5HOs in hibays

- Can be started down to –20°F with proper ballasting

- Minimize lamp and ballast types

Lamp minimization can be so important to facility and maintenance departments because fewer types have to be purchased, stocked, and carried around. If 28W and 25W F32T8s are used in offices, classrooms, etc., usually T5HO lamps and ballasts

Table 7-1

4' LINEAR FLUORESCENT EFFICACY TABLE

A	B	C	D	E	F	G	H	I	J	K	L	M	N
4' lamp type	initial catalog or photopic lamp lumens	lamp watts	lamp lumens per lamp watts	lamp quant	ballast type	standard ballast factor	system watts	initial system lumens	initial system lumens per watt	mean or 8000 hour lumen maintenance	mean or 8000 hour system lumens	mean or 8000 hour system lumens per watt	percent light compared to best
high performance F32T8	3100	32	96.9	2	HP IS	0.87	53	5394	101.8	95%	5124	96.7	100%
	3100	32	96.9	2	HP PS	1.15	70	7130	101.9	95%	6774	96.8	100%
	3100	32	96.9	2	G IS	0.87	58	5394	93.0	95%	5124	88.4	91%
extra long life 2950 lumen F32T8	2950	32	92.2	2	HP IS	0.87	53	5133	96.8	95%	4876	92.0	95%
	2950	32	92.2	2	G IS	0.87	58	5133	88.5	95%	4876	84.1	87%
basic grade F32T8	2800	32	87.5	2	HP IS	0.87	53	4872	91.9	95%	4628	87.3	90%
	2800	32	87.5	2	G IS	0.87	58	4872	84.0	95%	4628	79.8	82%
30W F32T8	2850	30	95.0	2	HP IS	0.87	51	4959	97.2	95%	4711	92.4	95%
	2850	30	95.0	2	G IS	0.87	55	4959	90.2	95%	4711	85.7	89%
28W F32T8	2750	28	98.2	2	HP IS	0.87	48	4785	99.7	95%	4546	94.7	98%
	2750	28	98.2	2	G IS	0.87	51	4785	93.8	95%	4546	89.1	92%
25W F32T8	2440	25	97.6	2	HP IS	0.87	42	4246	101.1	95%	4033	96.0	99%
	2440	25	97.6	2	G IS	0.87	47	4246	90.3	95%	4033	85.8	89%
extra long life 25W F32T8	2400	25	96.0	2	HP IS	0.87	42	4176	99.4	95%	3967	94.5	98%
	2400	25	96.0	2	G IS	0.87	47	4176	88.9	95%	3967	84.4	87%
high lumen F28T5	3050	28	108.9	2	HP PS	0.95	58	5795	99.9	93%	5389	92.9	96%
typical F28T5	2900	28	103.6	2	G PS	1.00	64	5800	90.6	93%	5394	84.3	87%
26W F28T5	2900	26	111.5	2	HP PS	0.95	55	5510	100.2	92%	5069	92.2	95%
49W F54T5HO	5000	49	102.0	2	HP PS	1.00	105	10000	95.2	93%	9300	88.6	92%
typical F54T5HO	5000	54	92.6	2	G PS	1.00	117	10000	85.5	93%	9300	79.5	82%
F34T12 800	3100	34	91.2	2	G RS E	0.85	60	5270	87.8	93%	4901	81.7	84%
F34T12 CW	2650	34	77.9	2	G RS M	0.88	72	4664	64.8	87%	4058	56.4	58%

notes: Lumens, lumen maintenance, ballast factors and wattages may vary among various manufacturers.

28W and 25W F32T8s, T5s and T5HOs have optimal light output at 92 - 95F compared to 77F for full wattage T8s.

Although efficacy can be improved with IS and RS ballasts with T5s and T5HOs, lamp life can be greatly reduced and lamp manufacturers may not warranty lamps.

93% is used as an average EOL lumen maintenance for T5HOs. 90% - 94% range among manufacturers.

All wattages based on 277V. HP is high performance. G is generic. IS is instant start. PS is program start. RS is rapid start. E is electronic. M is magnetic.

Extra long life is 36,000+ hours with IS and 40,000+ hours with PS ballasts at 3 hour cycles.

Prepared by Stan Walerczyk of Lighting Wizards   www.lightingwizards.com   1/1/14 version

have to be used in hibays because the reduced wattage T8s do not usually have sufficient lumens for hibays. Reduced wattage T8s should not be used below 60°F. So, the facility and maintenance people are stuck with two types of lamps and ballasts. On the other hand, high lumen 32W F32T8s can be used almost everywhere.

SPECTRALLY ENHANCED LIGHTING

If the reader is reading only selected chapters, Chapter 6, Spectrally Enhanced Lighting, should be read here, because it is very relevant to the current chapter.

WHY BETTER REBATES FOR 28W AND 25W F32T8s?

Numerous programs give higher rebates for 28W and 25W F32T8s than for high lumen 32W F32T8s, and some programs allow only the reduced wattage T8s, even though more energy can usually be saved with the high lumen 32W F32T8s. Often only 28W and 25W F32T8s are included in prescriptive rebate programs, and 32W F32T8s have to be used in customized rebate programs based on how much kWh is saved over the first year. Unless there are high annual hours of operation, the rebate is often much higher with the reduced wattage T8s, even if significantly less energy is saved. Some prescriptive rebate programs allow only up to 50% delamping, which provides a higher rebate for using more 28 or 25W F32T8s than for using fewer high lumen 32W F32T8s, even though there is sufficient light and more savings with fewer high lumen 32W F32T8s.

Except maybe for emerging technologies, rebate programs should be technology neutral and based on energy savings. I would appreciate any help getting rebate organizations to make their rebate programs fairer.

NO MAN'S LAND

Going lamp-for-lamp with 28W or 25W F32T8s can save energy, but often after they are installed there is not a sufficient amount of potential savings to go with an optimal solution, such as with delamping with high lumen 32W F32T8s and high performance ballast. Thus, maximum energy savings is sacrificed for partial energy savings in this "no man's land."

DELAMPING & LAMP-FOR-LAMP COMPARISONS

Following are just a few examples.

2x4 Lensed Troffer with 2 F34T12s or Basic Grade F32T8s

These troffers can be retrofitted lamp-for-lamp or delamped. Several companies have one-cove reflector kits. Consider Table 7-2; the bolded columns are the most important. Visually effective lumens are based on spectrally enhanced lighting. If maintaining existing light levels is important, the two gray background options are probably the best. However, since payback does not include any benefit after the payback period (and there usually is more benefit after payback), the long-term benefit column is much better. Net present value, modified internal rate of return, and savings to investment ratio are even better financial tools for evaluation.

2x4 Lensed Troffer with 3 or 4 F34T12s or Basic Grade F32T8s

These troffers are worse than two-lampers because these have lower efficiency since there are usually two lamps in one reflective compartment with nothing between the lamps, so light is wasted from bouncing directly between pairs of lamps and bouncing indirectly due to the lamp light hitting the reflective housing and then hitting another lamp. Unless these fixtures are air handlers, the heat from the lamps and ballasting can bring the ambient air temperature in the lamp compartment to about 100°F,

Table 7-2

2x4 lensed troffer with 2 basic grade F32T8 741s, 2-lamp .88 BF electronic generic ballast, angled interior sides with decent white interior housing & clear prismatic lens

$0.15	KWH rate	4000	annual hours	additional savings from reduced AC load
				1.1

Proposed — $0.05 KWH incentive — $0.00 — 10 cumulative years in long term benefit

Existing

type	watts	annual elect. cost	lamp life @ 3 hour cycles	mean regular lumens	mean visually effective lumens
2 basic grade T8 741 lamps, 2-lamp .88 BF generic electronic ballast, angled interior sides with decent white interior housing & clear prismatic lens	59	$35	20,000	4717	6651

Proposed

retrofit and relamping options	mean regular lumens	mean task mod. lumens	fixture mod-ifier	fixture mean regular lumens	fixture mean visually effective lumens	watts	watt reduc-tion	annual elect. savings	incen-tive	appr. installed cost	pay-back (yrs)	lamp life @ 3 hour cycles	compre-hensive long term benefit	notes
2 high lumen 32W F32T8 841s & 2-lamp .77 BF high performance ballast	4543	6619	1.05	4770	6950	48	11	$7.26	$2.20	$45	5.9	24,000	$37	
2 high lumen 32W F32T8 850s & 2-lamp .77 BF high performance ballast	4404	7415	1.05	4624	7786	48	11	$7.26	$2.20	$45	5.9	24,000	$37	
2 high lumen 32W F32T8 841s & 2-lamp .71 BF high performance program start ballast	4189	6103	1.06	4419	6439	46	13	$8.58	$2.60	$51	5.6	30,000	$55	
2 high lumen 32W F32T8 850s & 2-lamp .71 BF high performance program start ballast	4061	6837	1.06	4284	7213	46	13	$8.58	$2.60	$51	5.6	30,000	$55	
2 high lumen 32W F32T8 850s & 2-lamp .60 BF high performance program start ballast	3432	5778	1.06	3638	6125	44	15	$9.90	$3.00	$50	4.7	30,000	$72	
1 high lumen 32W F32T8 841, 1-lamp 1.20 BF high performance ballast & 1-cove white reflector	3540	5037	1.15	4071	5793	38	21	$13.86	$4.20	$60	4.0	24,000	$117	probably not enough light
1 high lumen 32W F32T8 850, 1-lamp 1.00 BF high performance ballast & 1-cove white reflector	2950	4956	1.16	3422	5749	33	26	$17.16	$5.20	$58	3.1	24.00	$162	probably not enough light
1 high lumen 32W F32T8 850, 1-lamp 1.20 BF high performance ballast & 1-cove white reflector kit	3540	5947	1.15	4071	6839	38	21	$13.86	$4.20	$60	4.0	24,000	$117	
1 high lumen 32W F32T8 850, 1-lamp 1.15 BF high performance program start ballast & 1-cove white reflector kit	3393	5700	1.15	3902	6555	39	20	$13.20	$4.00	$66	4.7	30,000	$116	
2 28W F32T8 841s & 2-lamp .77 high performance ballast	4004	5845	1.09	4364	6371	42	17	$11.22	$3.40	$45	3.7	24,000	$82	
2 28W F32T8 850s & 2-lamp .71 high performance program start ballast	3850	5964	1.09	4197	6501	40	19	$12.54	$3.80	$51	3.8	30,000	$103	
1 28W W F32T8 850, 1-lamp 1.15 BF high performance ballast & 1-cove white reflector kit	2990	5023	1.17	3498	5877	33	26	$17.16	$5.20	$60	3.2	24,000	$160	probably not enough light
2 25W F32T8 841s & 2-lamp .87 high performance ballast	3550	5894	1.09	3870	6424	42	17	$11.22	$3.40	$46	3.8	24,000	$81	
2 25W F32T8 850s & 2-lamp .77 high performance ballast	3465	5821	1.10	3812	6403	38	21	$13.86	$4.20	$46	3.0	24,000	$111	

footnotes: In excel version, numbers in colored boxes can be changed, which automatically alters computations.

Visually effective lumens are based on 'paper' tasks, which is $P(S/P)^{.78}$. Wattage is based on 277V.

An example of a con is 4 lamps per cross section which creates higher than optimal temperatures and wasted light from light bouncing in adjacent lamp. An example of a pro, which increases fixture efficiency. Another pro may be closer to optimal 77 degrees

3000-3100 lumen 32W F32T8 835/841/850 lamps are considered to have 80 - 86 CRI and 24,000 hour rating with instant start ballasts. They include Philips Advantage and GE HL (high lumen). Lamp life may be increased with .6 BF program start ballasts.

Additional air conditioning savings are because less watts translates directly into less heat. This can range for 1.0, which is no air conditioning savings, like in San Francisco to 1.15, which is an additional 15% savings in hot areas, like Sacramento.

Long term benefit is (annual electrical savings, including reduced air conditioning load x # of years) + incentive - initial cost. There are adders for delamping and 24,000 hour 32W T8s.

Long term year benefit does not include maintenance labor savings or cost of money, which can be considered to cancel each other out.

Approximate pricing is for medium sized projects. Pricing could be less for large projects. Pricing could be more for smaller, union and/or prevailing wage projects.

The 2x4 lensed troffers may not be able to be delamped easily if power comes in through one end in the center.

Fixture modifier is based on increasing or decreasing fixture efficiency and how close to optimal 77 degrees F for full wattage T8s in lamp compartment.

Copyright of Stan Walerczyk, LC, principal of Lighting Wizards. 1/1/14 version.

or even higher.

Delamping down to two better T8s and high performance ballast can bring the temperature down in non-air handler troffers much closer to the optimal 77°F for full wattage T8s. Delamping also only has one lamp in a each reflective chamber, which increases fixture efficiency.

If the long interior sides are angled and the original white paint or powder coating is in good shape, often a reflector kit is not necessary with delamping to two lamps. Often, the existing outboard lamp holders can be used and, for rebate purposes, the inner lamp holders can be removed. Although this will provide sufficient light, the lamps are usually not centered, which is okay for some end customers. Sometimes a higher BF is required without a reflector.

If the existing troffers do not have angled, long interior sides and if the reflective surfaces are not in good shape and cannot be cleaned, reflectors should be used with delamping. If the end customer wants centered lamps for aesthetics, a reflector or centering-bracket kit should be used.

2x2 Lensed Troffer with 2 U-bend T12s or T8s

Although every facility manager and maintenance person with whom I have communicated has wanted to get rid of U-bend lamps because of cost and bulkiness, however there are still some retrofit contractors and others that want to retrofit fixtures with full or reduced wattage U-bend T8s.

The best retrofit is usually a two-cove white reflector kit, two high lumen 17W F17T8s, and a .87-1.15 BF high performance ballast. The reflector is necessary to get power to both sides of the fixture. The BF depends on how much light is needed.

8' Strip with two 60W F96T12 CWs

Almost every facility manager and maintenance person wants to get rid of 8' lamps, but some retrofit contractors and others want to retrofit these fixtures lamp-for-lamp with full or reduced wattage 8' T8s.

A kit is necessary to convert this fixture with 4' T8s. If the existing up and side light is important, it can be a strip kit. To provide as much light as existing, the best solution is usually two high lumen 32W F32T8s and high performance 1.15-1.20 BF ballast, which reduces wattage from about 123 to 70-72. If the existing up and side light is really wasted light, then it can be a deep or shallow-hooded industrial reflector kit. Often, two high lumen 32W F32T8s and .71-.89 BF high performance ballasts work the best, maintaining existing light levels where needed and reducing wattage from 123 to 46-60.

8' Strip with 4 F34T12s or basic grade F32T8s
This is similar to the above, but another option is delamping without a kit and installing two lamps in opposite corners.

2x4 18-Cell Parabolic Troffer with 3 F34 T12s or Basic Grade F32T8s
Parabolic troffers were popular in the late 80s and early 90s, when computers became popular in offices. Those computers had curved screens without anti-glare coating. Parabolics helped, but now computer screens are flat and have anti-glare coating. The downsides of parabolic troffers are currently known:

- Dreaded cave effect from dark ceilings and upper walls, which makes the space seem dark and smaller

- 70-75% fixture efficiency

- Lack of vertical footcandles

- Overhead glare, which can cause eye strain and headaches

Figure 7-3 shows one example of an office worker hating parabolic troffers. The maintenance person told me that it took two months to get this office worker to take down the tent.

Some retrofitters and others go lamp-for-lamp with 28W or 25W F32T8s, but that keeps all of the given downsides of parabolics.

Figure 7-3. A drastic measure to combat ill effects of parabolic troffers

Some retrofitters and others go for two-cove white reflector and two high lumen 32W or reduced wattage F32T8s. Light levels can be okay, but for office applications, this ruins proper cut-off angles, which can create a glare bomb. Plus, if a specular reflector is used, the fixture looks like a house of mirrors.

Replacing parabolic louvers with upscale kits is a great way to do an upgrade (and usually a retrofit without an upgrade is a wasted opportunity). A.L.P Lighting, Energy Solutions International. Envirobrite, Harris Lighting, Lithonia, and others have upscale kits that greatly improve lighting quality and greatly reduce wattage in parabolic troffers. (See Figure 7-4.) Up to three F34T12s or basic grade T8s can be replaced with one high lumen 32W F32T8. If 5000 or higher Kelvin is used, the BF may be as low as .87. This kit can also be used with 28W or 25W F32T8s, T5s, and T5HOs.

Figure 7-4. An Energy Solutions International kit

REFLECTORS

Size Matters

Some customers want low prices, even if performance suffers. But with proper design, fewer lamps and/or lower BF can be used to save more energy, which is really a better solution.

One way some manufacturers try to keep costs down is to design most or all reflectors for smaller diameter T5s and T5HOs, and then also use those same reflectors with larger diameter T8s, which reduces efficiency and optical control with T8s.

ESI and some manufacturers have commodity-grade reflector kits for hibays and other fixtures, as well as more expensive spec-grade kits and fixtures that are usually a better total value because more wattage can be saved cost effectively. One example of a good T8 high bay is the ESI Premium HS Series.

I usually only recommend reflectors for T8s that are at least 4" nominally wide (which can be down to 3.75" wide), with the reflector not protruding farther than the lamp. Also, there should be only one lamp per reflector cove.

Materials

Over the years in typical offices, classrooms, and halls, specular reflectors have been more popular in the east, while white reflectors have been more popular in the west. For most of these applications, white is better because of less glare, better uniformity, and not seeing the reflector bends on the lens.

There are primarily two types of white reflectors, often called prepaint and white powder coat. Prepaint is usually less expensive and has slightly less reflectivity. White powder coat can be considered more environmentally friendly because no paint thinners or solvents are used.

For high heights, specular is usually the best. For mid-heights in gyms, retail stores, and conference centers, often the best material is Alanod's Micro Matt, which looks white for less glare but has the optical control of specular.

MANUFACTURERS

Some manufacturers of reflector kits and lamp holder kits, as well as fluorescent hibay and other fixture manufacturers that you may already know about include:

- A.L.P. Lighting
- Amerillum
- Energy Solutions International
- Envirobrite
- Harris Lighting
- Precision-Paragon

Lithonia, Daybrite, and some other large manufacturers have also gotten into kits because the new construction market has not been very good.

Chapter 8

Induction

This chapter is short, but no matter what you have read or heard before, please don't specify or buy any induction product, except maybe for some strange application.

Induction has mediocre LPW, including lamp and generator. Fixture efficiency and optical control are typically bad because the lamps are so large. I have not considered induction since the fall of 2009 and do not understand why it is still rebatable, because the lamp and generator LPW is in the same ballpark as common-type T12 lamps and ballasts, and those lamps have been eliminated. Induction is a mature technology, which is getting further behind LED and electronically ballasted CMH each and every year.

The question is, if somebody buys and installs induction now, will the manufacturer still make induction lamps and generators or even be in business down the road to handle warranty and other issues? A few years ago, Philips, the first manufacturer of induction, sold its entire line to focus on high performance technologies, including fluorescent, CMH, and LED. Table 8-1 shows how bad induction is.

In 2014 and beyond, LED and CMH will get better, while induction and fluorescent will stay about the same.

(A footnote is that while writing this, I received another email from an unknown Chinese induction manufacturer. I wish all of those would automatically go into my spam folder!)

Table 8-1

HOW DOES INDUCTION COMPARE IN 2014?

technology	initial lamp & ballast, driver or generator LPW	typical fixture efficiency	initial fixture LPW	EOL lumen maintenance	EOL fixture LPW	optical control	lamp & ballast, driver or generator lives
induction lamp & electronic generator	72 - 88	65 - 80%	47 - 70	65 - 70%	31 - 49	large lamp, so bad	100,000 - 65,000 - 100,000
LED & electronic ballast			80 - 120	70%	56 - 84	very good, so less LPW okay	50,000 - 100,000+ for both
CMH lamp & electronic ballast	110 - 115	75 - 85%	83 - 98	80%	66 - 78	small lamps help	30,000 60,000
F32T8 lamp & electronic ballast	100 - 102	75 - 90%	75 - 92	90%	68 - 83	often not the best	24,000 - 67,000 60,000

LPW = lumens per watt   EOL = end of life   CMH = ceramic metal halide

Chapter 9

Electronically Ballasted CMH

INTRODUCTION

Everybody knows about LED, and some people mistakenly consider induction good, but many people do not know how good electronically ballasted ceramic metal halide (CMH) really is.

CMH

Although CMH is quite good with magnetic ballasts, it usually takes electronic ballasts to allow these lamps to compete with LED. Electronically ballasted CHM can provide 110-120 LPW and up to 95 CRI. Lamp life can be 30,000-50,000 hours.

The time required to replace lamps can be doubled by connecting one 1-lamp electronic ballast to two lamps so that 30,000-50,000 hours can turn into 60,000-100,000 hours. Each time power is turned on, the lamp with the least resistance turns on. So, it could be lamp A on day one, lamp B on day two, lamp A on day three, etc. Figure 9-1 shows an example.

Of commonly available lighting sources, only LED and electronically ballasted CMH are significantly improving. At least one manufacturer is working on developing reduced wattage lamps with the same lumens as existing lamps that can work with existing ballasts. For example, if a 90W CMH lamp and ballast is used now, when that lamp burns out, there may be by then about

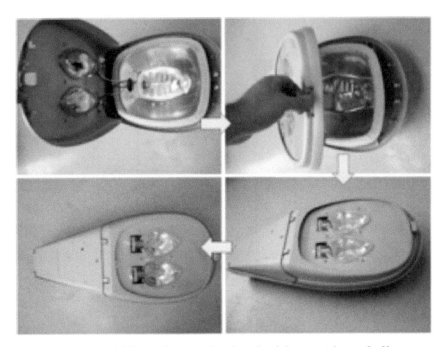

Figure 9-1. Philips 2-lamp cobrahead with one 1-lamp ballast.

a 75W CMH lamp available that can be used as a straight replacement without changing the ballast.

Philips can probably be considered the major player with electronically ballasted CMH, especially with its long history in Europe. Philips has Cosmopolis and Elite lamp and ballast systems. Often it is good to get lamps and ballasts from the same manufacturer so that if there is any problem, manufacturers cannot point blame fingers at each other.

There are several other manufacturers that either use the Philips lamps with the Philips label or private label them and make their own electronic ballasts. Ventura Lighting's Rio system includes its own reflector amp holder reflector.

Philips also has the AllStart CMH lamp, which is designed to work with probe and pulse start magnetic ballasts. These lamps save energy, improve CRI, and have longer life than standard MH lamps. Installed costs are relatively low because there is no new ballast to buy and install, but there is not as much electrical reduction when keeping a magnetic ballast than when using an electronic ballast.

Most LED hibay and exterior fixtures have only about 70 CRI. Where CRI is really important, such as with big box stores and car dealerships, 90+ CRI CMH can make colors and textures look better, which can make the merchandise more sellable.

COMPARISON TABLE

Architectural grade exterior fixtures, typically cast metal, are expensive no matter the light source. If the fixtures, reflectors, and lenses are in good shape, it is often much more cost effective to retrofit them with electronically ballasted CMH than to replace them with new, similar-grade LED fixtures. (See Table 9-1.)

Table 9-1

ARCHITECTURAL GRADE LIGHTING FIXTURE FEASIBILITY STUDY

$0.150	blended KWH rate	$0.05	/KWH saved over first year rebate										15					
																	long term benefit years	
existing						proposed												
fixture type	rated lamp life at 10 hour cycles	watts	annual hours	tech-nology	option	retrofit or replacement fixture option description (poles, arms, etc. not included)	rated life	watts	watts reduc-tion	% watts reduc-tion	annual electrical savings	appr. installed cost	rebate	long term benefit just electricity	per year maintenance savings, improved lighting quality, etc. benefit for comprehensive long term benefit and payback	long term benefit compre-hensive	payback in years just electricity	payback in years compre-hensive
exterior architectural grade fixture with 150W HPS lamp & magnetic ballast	24,000	188	4200	EHID	A	1 60W CMH lamp, maybe lamp holder adapter & electronic ballast	30,000 - 50,000	68	120	64%	$75.60	$250	$25.20	$884	$20	**$1,209**	3.0	2.4
					B	1 90W CMH lamp, maybe lamp holder adapter & electronic ballast	30,000 - 50,000	99	89	47%	$56.07	$250	$18.69	$591	$20	**$910**	4.1	3.0
					C	2 60W CMH lamps, custom lamp holder kit & 1-lamp electronic ballast (only lamp on a time)	60,000 - 100,000	68	120	64%	$75.60	$350	$25.20	$784	$35	**$1,334**	4.3	2.9
					D	2 60W CMH lamps, custom lamp holder kit & 1-lamp electronic ballast (only lamp on a time)	60,000 - 100,000	99	89	47%	$56.07	$350	$18.69	$491	$35	**$1,035**	5.9	3.6
				LED	E	new similar grade architectural grade LED fixture matching existing green paint with 10 year warranty	100,000	65	123	65%	$77.49	$1,000	$25.83	$162	$45	**$863**	12.6	8.0
					F	new similar grade architectural grade LED fixture matching existing green paint with 10 year warranty	100,000	90	98	52%	$61.74	$1,200	$20.58	-$274	$45	**$422**	19.1	11.0

prepared by Stan Walercyk of Lighting Wizards 1/1/14

Chapter 10

Important Stuff to Know about LEDs

INTRODUCTION

This chapter discusses some particularities of LEDs and relationships to other lighting technologies.

DOE

The DOE has a ton of information on LEDs. Although the full website address is quite long, you can Google search "DOE SSL" for the Department of Energy Solid State Lighting's website. Some of what you can find includes:

• CALiPER test reports

• Case studies

• Benchmark reports

• Gateway studies

• Fact sheets

• Short webinars

If you enter your name and email address, the DOE will inform you of anything new.

LUMENS

Most incumbent lamps are standardized and have bare lamp lumens with or without ballasts. There really are no standardized LED packages; lumens are rated out of the complete product or fixture. So, comparing incumbent technologies and LEDs should be apples to apples, out of the product or fixture. For example, two 3100-lumen high performance fluorescent T8 lamps and .87 BF ballast provide 5394 bare lamp lumens, but putting them in a typical clear prismatic lensed troffer with 85% fixture efficiency and 5% thermal losses, there may only be 4315 out of fixture lumens. That can be compared to an LED troffer that provides about the same out of fixture lumens.

KELVIN

The most common white LEDs are really blue LEDs with phosphors, similar to what is used in fluorescent lamps. The less that blue light has to be transformed, the higher the lumens and LPW. So, 6000K LEDs are generally more efficient than 5000K, which are more efficient than 4000K, which are more efficient than 3500K, which are more efficient than 3000K. Sometimes the CRI is reduced to minimize lumen drop with lower Kelvin LEDs. Although 5000K and 6000K LEDs may be okay for exterior applications, people often want warmer color tones for interiors, but 5000K and higher CCT LEDs and other light sources can be good for human centric lighting.

LUMENS PER WATT

This is similar to lumens. Even though the fluorescent T8 system provides 100 LPW bare lamp with ballast, there may only be 80 LPW out a lensed troffer. Thus, an LED troffer with 90 LPW out of the fixture is better. There some LED troffers and troffer kits with 100 and even 110 LPW.

COLOR RENDERING

CRI has been used for decades and has worked quite well with incumbent lighting sources. It is based on eight pastel colors. Daylight and incandescents are considered to have a perfect 100 CRI. On the other end, LPS has 0 CRI. Everything else is in between. But CRI does not work well with most LEDs because LEDs do better with bright (saturated) colors than with pastel (unsaturated) colors.

Most interior LED products have about an 80 CRI, but there are some in the 90s. Most exterior LED products and hibays have about a 70 CRI. So if you want to highlight bright colors, LEDs, even with lower CRI than other products, may be best.

Color quality scale (CQS) has both bright and pastel colors and may replace CRI in the future.

R9

R9 is red and is considered the ninth color cell. Many LEDs have little very red content, so skin tones and red objects look washed out. Some LEDs even have a negative R9 value. For many applications, such as a grocery store freezer with red meat, it is important to get LEDs with a relatively high R9 value.

OPTICAL CONTROL

Well-designed LED products can have excellent optical control, providing the right amount of light where it should go and not wasting light where it is not useful. Plus, they can avoid that big blob of excess light underneath that other technologies produce. Often well-designed LED products with significantly less total lumens can outperform other high performance technology products.

Even if LED products with great optical control have less total lumens than other products, well-designed LED products can work because they can provide the light where necessary without providing light where it is not useful, such as the big underneath blob of light or light in front of the fixture. Some LED products do not have good optical control and are okay when that is not important.

GLARE

Shining light where it should go and where it should not go makes glare a likely suspect.

L70—LIFE

Incumbent lamps' rated life is based on when half of them have burned out and half are still working. Starting is the hardest part on fluorescent and HID lamp life; it would actually be better to rate life on number of starts.

LEDs are different. Life is not shortened with short cycles, and they do not really die. Instead, they get dimmer and dimmer over time. Rated life is L70, which is when LEDs still have 70% of initial lumens. As soon as they lose more than 30%, they are considered past rated life. But the life of LEDs is usually not the weakest link. Drivers, electrical connections, hinges, etc. could fail before LEDs. Many manufacturers list 100,000 hours or longer rated life but offer only a five-year warranty, so they must not feel comfortable that the drivers and other components will last that long. I consider warranty length much more important than L70.

Also, LED life is maintained or improved with turning on and off because they get cooler when turned off. Fluorescent and HID lamp life gets worse when being turned on and off, manually or automatically. This can be important with many occupancy sensor applications.

LM79

This is probably the most important report on LED products because it includes lumens, LPW, CRI, etc. LM79 reports should only be done by labs approved by the DOE. Do not buy any LED product that does not have an LM79 report from a good lab. Also, do not buy an LED product unless the performance is good compared to similar LED products.

DIMMING

All LED chips can dim, but not all LED products can dim—and even ones that can cannot dim with all dimmers. Many LED products do not work with various incandescent phase-cut or wave-chopping dimmers. Often, various LED products need to be tested with existing incandescent dimmers, or the incandescent dimmers have to be replaced with ones that have been designed for LEDs.

LED systems can maintain or improve LPW while dimming because they run cooler when dimmed. This is much better than fluorescent dimming, where efficiency gets less because the ballasts have to heat the cathodes of lamps below .70 BF. This heating consumes wattage, and with LED the lamps do not flicker, spiral, or turn off.

FLICKER

Many people think that flicker vanished with fluorescent magnetic ballasts, riding off into the sunset. But flicker has raised its ugly head again with many LED products. Some LED products flicker when fully on, and flickering gets worse when dimmed. Other LED products do not flicker when fully on but flicker when dimmed, especially deeply dimmed.

There are two types of flicker—visual and non-visual but sensed. Both can cause eyestrain, headaches, reduced acuity, and even seizure. When you can see flicker, that is a useful signal to avoid it. But what can you do if you do not see the flicker but are aware that there may be a problem? Just use your cell phone camera. Before taking a picture, check to see if you see any bars on the screen; if you do, there is flicker. This can also be used to check magnetic fluorescent ballasts.

Hopefully, the DOE and others will develop approved flicker metrics and levels.

HEAT

Heat is not good for LEDs and drivers, so it is important to keep them cool as cost effectively as possible. In anything other than in a refrigerator or freezer, make sure the LED products that you specify or buy have good thermal dissipation for both LEDs and drivers. This can be done in various ways, including with thick metal fins, thin metal fins, fans, and modified radiators.

Check the highest ambient temperature rating for LED products. One caution is that many screw-in LED lamps are not supposed to go into small enclosed fixtures, such as jelly jars and pagodas. Another is that some LED hibays are not supposed to be used above 120-125°F, and at fixture height many warehouses and industrial facilities get much hotter than that during hot summer days.

WARRANTY

First of all, exceeding ambient temperature rating voids warranty. It is very important to read warranties from various manufacturers, even if the length of the warranty is the same. There can be big differences in coverage. Although most fluorescent and HID ballasts have a parts and a labor warranty, most LED products only have a parts warranty.

The standard length of warranty on many LED products is five years. Several manufacturers will double that to ten years if the customer is willing to pay an extra 5-10% up front, which is often a very good value for the customer. A ten-year warranty is often required, because with a five-year one, the payback happens after warranty expires, which does not enthuse end customers. The warranty should be longer than the payback period.

But how valid is the warranty? It may only be worth the paper it is printed on from a new or small manufacturer. Even some long established, top tier manufacturers may not be doing well in five or ten years. Some of them may be even be out of business, due to how fast LEDs and other SSL is progressing and changing directions.

Accordingly, manufacturers or sales companies backing up warranties with good third-party insurance policies may be the best solution. This has already started to happen. I consider a good warranty, especially if backed up in this manner, much more important than rated life. So what if a manufacturer lists 100,000, 150,000, or even 200,000 hours of rated life, if they only provide a five-year warranty? If they really thought their products were that good, they would provide a ten-year warranty.

Chapter 11

Store Cooler Lights

This is a short chapter on a relatively limited, but good opportunity.

LED kits and fixtures can be great solutions in refrigerators/ freezers with vertical doors, and more are being developed for the open, horizontal, reach-down models. Usually more energy can be saved with reduced cooling load than from reduced lighting load, because LEDs do not have any heat on the light side.

As a vegetarian I couldn't care less, but for omnivores who want to buy red meat, it should look red. Some LEDs do not have R9 content, which means that red meat and other red products appear washed out and not that appealing.

Occupancy sensors, especially those that dim up and down, can often be very cost effective, especially in stores open all night, with few customers in the middle of the night. (See Figure 11-1.) There is some anecdotal evidence that sales may go up when the lights get brighter as people walk toward a refrigerator or freezer. For example, think about how it would be walking toward ice cream, with the lights getting brighter as you get closer.

Figure 11-1. Occupancy sensors in supermarket lighting

Chapter 12

Decorative Lights

INTRODUCTION

LEDs are very good for these applications. Make sure you get at least a three-year warranty.

CHRISTMAS/DECORATIVE LIGHTS

Although most people only use LEDs for Christmas, numerous restaurants, bars, and hotels use them all year long for lighting, both interior and exterior. There is almost no reason not to use LED.

CHANDELIERS

These fixtures include cut glass and other types that have medium, intermediate, or miniature screw-in bases. The amount of ambient light is often not that important, but making the fixtures look good is. Three important factors are:

- Long life (because it is usually not fun getting on a ladder getting to these lamps)

- Energy efficiency (low wattage)

- Sparkle (to high-light the gut glass and/or other parts of the fixture)

Clear incandescent bulbs have sparkle, but not long life or energy efficiency. CFLs have long life and energy efficiency, but with the phosphor on the glass, no sparkle. LEDs have all three. Figure 12-1 illustrates an example of each.

Figure 12-1. CFL bulb (left) and LED bulb (right)

Chapter 13

Omnidirectional Lights

INTRODUCTION

Omnidirectional is a fancy term for screw-in lamps that provide light in all directions, such as typical incandescent A19 and spring CFL medium-base, screw-in lamps. Although CFLs have been the best value, especially with upstream rebates, some good LEDs became available for $10 or less in mid-2013. So a shift has really begun.

INCANDESCENTS

The United States and other governments have been eliminating standard incandescent bulbs, but there are available energy saving halogen and halogen-infrareds that turn on, turn off, and dim just like standard incandescents. CRI is 100 for all of these. Rated life is quite short.

CFL

Screw-in CFLs can be as low as $.25 with upstream rebates, but even if costing up to $5, they can be quite cost effective regarding lumens, LPW, and life. Recent ones are pretty darn good with instant on and about 80 CRI and 6,000-10,000 hours rated life. Those that state dimmable can usually dim down to about 20% and can work with many, but not all, existing incandescent dimmers. CFLs tend to turn bluish or grayish with deep dimming, which does not look very good in restaurants and other applications. Most electronically ballasted screw-in CFLs should not be used inside small enclosed fixtures, such as recessed cans with lens and no venting, jelly jars, and pagodas.

What do you think has more mercury, a typical CFL or a can of tuna fish? Well, it can be about the same, or the tuna fish can have more. CFLs are not perfect, but they offer a good interim solution if people want to wait for LEDs.

LED

Screw-in LED lamps really need to be $10 or less and last at least 25,000 hours to be cost effective for many applications, but when maintenance costs are high, even $20 ones can be a good solution. Even $40 ones, which can replace 100W incandescents or 23W CFLs, may be cost effective in certain applications.

While it takes 13-15W CFLs to replace 60W incandescents, currently (2013) about a 10W LED can do it. In the future, the wattage will be less. Most of these are designed to dim, but not all of them dim with all incandescent dimmers, which are usually sine wave choppers. You may have to try various LED lamps with existing dimmers, keep at least one incandescent or halogen, or replace the dimmers with ones designed for LEDs. Some people want LED lamps that turn a warm color tone when dimmed, like incandescents do. There are already some of these available, and more will be developed.

Just like CFLs, most screw-in LEDs are not designed to go into small enclosed fixtures because of heat build-up. For these applications, check various manufacturers and models. Figure 13-1 shows an example of an of an LED lamp.

The new generation of dimming and color-changing screw-in LEDs, controlled by cell phone apps, began in 2013. Philips had one of the first products. (See Figure 13-2.) Now there are more, and there will be more in the future.

Figure 13-1. LED lamp from Cree

Figure 13-2. The Hue from Phillips

Chapter 14

Accent Lights

INTRODUCTION

Accent lights are reflector lamps that shine light in one direction. Halogen has been the standard, with a transition to halogen infrared, which is like halogen on steroids, but legal. There are CFL reflector lamps, which are fine when sparkle and focus are not important.

With pricing coming down, decent electric rates, decent rebates, and significant annual hours of operation, it is becoming an LED world for accent lighting. One big advantage of LED accent lighting is that there is no heat from light side, so produce, chocolate, flowers, etc. can usually last longer when lit by them. Without any infrared or ultraviolet energy, LEDs can be very good in museums, art galleries, etc.

Many LED reflector lamps can be quite glary in many applications. Where glare is an issue, LED reflector lamps with indirect lighting or with a high performance, milky lens may be good.

MR16 LAMPS

Hopefully, by the time this book is published, there will be LED MR16s, which can really replace 50W standard halogen and 35-38W halogen infrared MR16s, for one example. Even with less lumens, LED MR16s can often be used, especially if the area is currently overlit. When not overlit, additional track heads could be installed if necessary.

Most all MR16s are low voltage, which requires a step down transformer. Many older step down transformers do not work with LED MR16s because there is not enough wattage to activate them. Here are some solutions:

- Get various LED MR16s from various manufacturers, because some may work with existing transformers.

- If one transformer drives several MR16s, then often keeping one halogen or halogen infrared MR16 in the circuit will work.

- Get new transformers that are designed for LED MR16s.

- If using track lighting, get rid of transformers and MR16 track heads and get new line voltage track heads; use LED PAR20s, PAR30s, or PAR38s.

If you or your client still has standard halogen MR16s, halogen infrared ones could be used as an interim technology before LEDs are used.

PAR AND R LAMPS

These may cost $20-$50. Sometimes the higher priced lamps are worth it. Some rebate providers have upstream rebates through big box stores and/or distributors. With no heat from the light side, these and LED MR16s can extend the life of produce, flowers, etc. Dimming can be an issue with some LED lamps with some dimmers. Manufacturers have various cooling strategies, including:

- Thick metal cooling fins
- Thin metal cooling fins
- Computer-type fans

Figure 14-1 shows an example an LED PAR38, the MSi iPAR38. It can have three wattage and lumen settings, by inserting the ring in different ways.

Figure 14-1. MSi iPar38 lamp

No matter the cooling strategy, try to get ones with at least a 50,000-hour rating and five-year warranty. There are some exterior, rated ones, which can be used for lighting flags, signs, etc.

COMPARISON TABLE

Table 15-1 compares halogen infrared, LED, and CFL options. Please note that, although LED can be considered the best, some customers will not have the money for a large quantity of them. However, they probably have enough money for the other two options. Reflector CFLs should probably not be used to highlight merchandise or artwork because they do not have any sparkle or focus.

Table 14-1

TRACK LIGHT FEASIBILITY STUDY

$0.150	blended KWH rate	$0.05	/KWH saved over first year rebate											15		long term benefit years		
existing						proposed										long term benefit years		
fixture type	rated lamp life	watts	annual hours	tech-nology	option	retrofit or replacement fixture option description	rated life	watts	watts reduction	% watts reduction	annual electrical savings	appr. installed cost	rebate	long term benefit just electricity	per year maintenance savings, improved lighting quality, etc. benefit for comprehensive long term benefit and payback	long term benefit compre-hensive	payback in years just electricity	payback in years compre-hensive
track head with 90W standard halogen PAR38	6,000	90	5000	LED	A	well designed 20W LED PAR38	25,000 - 50,000	20	70	78%	$52.50	$65	$17.50	$723	$6	$830	0.9	0.8
				halogen infrared	B	60W halogen infrared PAR38	6,000	60	30	33%	$22.50	$10	$7.50	$328	$0	$335	0.1	0.1
				CFL	C	20W R40 CFL (NO SPARKLE & NO FOCUS)	10,000	20	70	78%	$52.50	$15	$17.50	$773	$2	$820	0.0	0.0

prepared by Stan Walercyk of Lighting Wizards 1/1/14

Chapter 15

Recessed Cans

INTRODUCTION

This is another very good application for LED. Most existing recessed cans can be retrofitted with LED PAR lamps or LED kits, but do not give up on CFLs in certain applications. That said, most facility and maintenance people that I have talked with want to get out of CFLs because rated life may only be 10,000 hours, meaning they have to replace and recycle lamps about every 1-1/4 to 1-1/2 years.

This chapter focuses on retrofitting existing recessed cans. When new ones are needed for new construction or a remodel, it is highly recommended to get generic cans and use LED PAR lamps or LED kits. Certain codes may require current limiters.

EXISTING 120V SCREW-IN

With existing screw-in incandescent or halogen, it is very easy to switch to screw-in LED. If the existing lamp is horizontal then an omnidirectional CFL or LED could be used. If the existing lamp is vertical, base up, then a reflector, PAR CFL, or LED lamp is good. If LED is chosen, it's often best to use the largest R or PAR lamp that will properly fit, because larger diameter LED lamps typically have more surface area and metal, so heat can dissipate better. Sometimes drilling holes in the top of recessed cans can also help reduce heat, but be mindful of UL and electrical codes.

No matter the existing lamp orientation, one of many LED kits could be used with existing screw-in base, but sometimes that base has to be detached from the housing and become floating. There are many manufacturers and models. Cree was the pioneer and now has numerous models. Figure 15-1 shows the remodeled LR6, which is available with several lumen packages in 120 and 277V.

EXISTING 120V PIN BASED CFL

Two-pin CFLs use magnetic ballasts, and four-pin CFLs use electronic
ballasts. Often magnetic ballasts only
drive one lamp. Some electronic ballasts are only designed to drive one or
two lamps, but some two-lamp electronic ballasts can be wired to drive
one lamp.

Figure 15-1. LR6

If a recessed can has two lamps, you can often delamp down
to one lamp. Some people think that will cut light levels in half,
but light levels are usually only cut by about one fourth because
fixture efficiency improves and lamps may operate at closer to
their optimal temperature. Higher Kelvin lamps may help, based
on the benefits of spectrally enhanced lighting. If it fits in the can,
a medium-base socket could be installed in top of the can, and an
LED R or PAR lamp could be screwed in. It is important to check
to see if this is a UL violation. One of the various LED kits could
be also be used with hardwiring.

EXISTING 277V PIN BASED CFL

This is similar to the above, but if LED R or PAR lamps are
used, a GU24 base is needed. If 120V LED lamps are used, a step
down transformer will also be needed. There are some 277V LED
lamps, but they are not that common. As above, check to see if
this is okay with UL. There are many 277V kits that are hardwired.

LED KITS ISSUES

Some of the kits may only work in vertical-sided recessed cans. Many manufacturers offer 4″ and 6″ kits, but they do not work in 8″ diameter recessed cans; however, Cree and some manufacturers have adapter rings for 8″ cans.

NEW RECESSED CANS

Although there are many dedicated new LED recessed cans, they are usually not recommended. Even if they are dimmable, the end user is stuck with CCT and light distribution. Plus, it may be difficult or virtually impossible to get replacement LEDs and drivers down the road. Thus, generic cans with either LED R or PAR lamps, or LED kits, are often recommended. If there are any power density code issues, current limiters can be used with LED lamps.

The Lighting Science Group 4″ and 6″ LED Glimpse can be used in recessed cans or just j-boxes, which can reduce cost. Figure 15-2 shows the 6″ version. Maybe other manufacturers have or will develop something similar.

COMPARISON TABLE

Figure 15-2. 6″ LED Glimpse

Table 7 is similar to the accent lighting table. At least for a recessed can with two two-pin CFLs and two one-lamp ballasts, it is much less expensive delamping than going with LED, but maintenance with 10,000 hour rated CFLs can be a concern.

Table 15-1

RECESSED CAN FEASIBILITY STUDY

$0.150	blended KWH rate	$0.05		/KWH saved over first year payback									10			long term benefit years		
existing					proposed													
fixture type	rated lamp life	watts	annual hours	tech-nology	option	retrofit or replacement fixture option description	rated life	watts	watts reduc-tion	% watts reduc-tion	annual electrical savings	appr. installed cost	rebate	long term benefit just electricity	per year maintenance savings, improved lighting quality, etc. benefit for comprehensive long term benefit and payback	long term benefit compre-hensive	payback in years just electricity	payback in years compre-hensive
recessed can with specular reflector, 2 horizontal 2700 - 3500K PL13 2-pin CFLs & 2 magnetic ballasts	10,000	34	5000	CFL	A	delamp with 1 4000 - 5000K PL13 in right or left side	10,000	17	17	50%	$12.75	$10	$4.25	$118	$2	$142	0.5	0.4
				LED	B	10W LED recessed can kit	50,000	10	24	71%	$18.00	$80	$6.00	$100	$8	$186	4.1	2.8

prepared by Stan Walercyk of Lighting Wizards 1/1/14

Chapter 16

LED T8s

INTRODUCTION

Due to safety and other concerns, LED T8s should be avoided. High performance fluorescent, LED light bars that do not use existing lamp holders, and hardwired LED kits with their own lenses are so much better.

GENERAL

Since there are hordes of sales people across the country pushing numerous brands of LED T8 lamps, many end customers think these products must be credible. But, high lumen or extra long life, full wattage fluorescent T8 lamps with high performance electronic ballasts, and the new generation of hardwired LED troffer kits are so much better with regard to cost, lumen maintenance, lamp life, cost effective energy savings, distribution, and safety.

LED T8s may cost $40-$60. Even with the recent price hikes, high lumen or extra long life, full wattage fluorescent T8s may cost $2.50-$6.00, depending on quantity and other factors.

Good fluorescent T8s only lose 8-10% of initial lumens by end of rated life, compared to 30% for LED T8s. So, what may be sufficient light initially with LED T8s may be underlit near the end of rated life.

High lumen fluorescent T8s are rated up to 42,000 hours with program start ballasts @ 12 hour cycles, and extra long life fluorescent T8s are rated up to 67,000 hours the same way. Most

LED T8s are rated for 50,000 hours, but that may be suspect with the heat in enclosed fixtures. High lumen or extra long life, full wattage fluorescent T8s and high performance electronic ballasting, with or without a reflector kit, are much more cost effective than LED T8s when replacing T12s and magnetic ballasts or basic grade T8s and generic electronic ballasts.

Good fluorescent systems are better than LED T8s. Most rebate programs, for good reasons, do not rebate LED T8s. Now that production of most T12 lamps has ended, rebates on high performance T8 systems to replace them may be based on the wattage of equivalent basic grade T8s and generic electronic ballasts.

The DOE and others have found that many LED T8s shine most of the light straight down, which can provide sufficient light there but not enough light between fixtures and on walls. LED T8s are terrible in parabolic troffers because, with no uplight, the top of the fixtures are dark. Some LED T8s do provide more sidelight, which can be good. Fluorescent T8s shine light in all directions.

Installing LED T8s, especially when keeping the existing fluorescent type of lamp holders, may void the UL listing of a fixture that was UL listed as a fluorescent fixture. If that is the case and something bad happens, like a fire, the insurance policy may not cover damages.

4 TYPES OF LED T8s

With Existing Magnetic or Electronic Ballast
It is as simple as removing existing fluorescent lamps and installing these LED T8s. But the existing ballasts, especially if they are magnetic, consume significant wattage. Plus, who knows how long the existing ballasts will last?

Table 16-1

4' LINEAR FLUORESCENT EFFICACY TABLE

A	B	C	D	E	F	G	H	I	J	K	L	M	N
4' lamp type	initial catalog or photopic lamp lumens	lamp watts	lamp lumens per lamp watts	lamp quant	ballast type	standard ballast factor	system watts	initial system lumens	initial system lumens per watt	mean or 8000 hour lumen maintenance	mean or 8000 hour system lumens	mean or 8000 hour system lumens per watt	percent light compared to best
high performance F32T8	3100	32	96.9	2	HP IS	0.87	53	5394	101.8	95%	5124	96.7	100%
	3100	32	96.9	2	HP PS	1.15	70	7130	101.9	95%	6774	96.8	100%
extra long life 2950 lumen F32T8	3100	32	96.9	2	G IS	0.87	58	5394	93.0	95%	5124	88.4	91%
	2950	32	92.2	2	HP IS	0.87	53	5133	96.8	95%	4876	92.0	95%
	2950	32	92.2	2	G IS	0.87	58	5133	88.5	95%	4876	84.1	87%
basic grade F32T8	2800	32	87.5	2	HP IS	0.87	53	4872	91.9	95%	4628	87.3	90%
	2800	32	87.5	2	G IS	0.87	58	4872	84.0	95%	4628	79.8	82%
30W F32T8	2850	30	95.0	2	HP IS	0.87	51	4959	97.2	95%	4711	92.4	95%
	2850	30	95.0	2	G IS	0.87	55	4959	90.2	95%	4711	85.7	89%
28W F32T8	2750	28	98.2	2	HP IS	0.87	48	4785	99.7	95%	4546	94.7	98%
	2750	28	98.2	2	G IS	0.87	51	4785	93.8	95%	4546	89.1	92%
25W F32T8	2440	25	97.6	2	HP IS	0.87	42	4246	101.1	95%	4033	96.0	99%
	2440	25	97.6	2	G IS	0.87	47	4246	90.3	95%	4033	85.8	89%
extra long life 25W F32T8	2400	25	96.0	2	HP IS	0.87	42	4176	99.4	95%	3967	94.5	98%
	2400	25	96.0	2	G IS	0.87	47	4176	88.9	95%	3967	84.4	87%
high lumen F28T5	3050	28	108.9	2	HP PS	0.95	58	5795	99.9	93%	5389	92.9	96%
typical F28T5	2900	28	103.6	2	G PS	1.00	64	5800	90.6	93%	5394	84.3	87%
26W F28T5	2900	26	111.5	2	HP PS	0.95	55	5510	100.2	92%	5069	92.2	95%
49W F54T5HO	5000	49	102.0	2	HP PS	1.00	105	10000	95.2	93%	9300	88.6	92%
typical F54T5HO	5000	54	92.6	2	G PS	1.00	117	10000	85.5	93%	9300	79.5	82%
F34T12 800	3100	34	91.2	2	G RS E	0.85	60	5270	87.8	93%	4901	81.7	84%
F34T12 CW	2650	34	77.9	2	G RS M	0.88	72	4664	64.8	87%	4058	56.4	58%

notes: Lumens, lumen maintenance, ballast factors and wattages may vary among various manufacturers.
In enclosed fixtures, since reduced wattage F32T8s consume less heat they can often operate closer to optimal 77 degrees F temperature, so may provide more light than this table shows compared to full wattage.
Although efficacy can be improved with IS and RS ballasts with T5s and T5HOs, lamp life can be greatly reduced and lamp manufacturers may not warranty lamps.
93% is used as an average EOL lumen maintenance for T5HOs. 90% - 94% range among manufacturers.
All wattages based on 277V. HP is high performance. G is generic. IS is instant start. PS is program start. RS is rapid start. E is electronic. M is magnetic.
Extra long life is 36,000 hours with IS and 40,000 hours with PS ballasts at 3 hour cycles.
Prepared by Stan Walerczyk of Lighting Wizards   www.lightingwizards.com 1/1/14 version

Table 16-2

4' T8 LAMP LIFE, LUMENS, CRI & MERCURY

LAMP	WATTS	3000-4100K LUMENS	3000-4100K CRI	5000K LUMENS	5000K CRI	MAX MG OF HG	LAMP LIFE HOURS INSTANT START 3 HR	INSTANT START 12 HR	PROGRAM START 3 HR	PROGRAM START 12 HR
1st GENERATION - GENERIC	32	2800	75-78	2800	75-78	1.7 - <10	15,000 - 24,000	20,000 - 30,000	20,000 - 30,000	24,000 - 36,000
2nd GENERATION - GENERIC	32	2950	81-85	2800 - 2950	80-85	1.7 - <10	15,000 - 24,000	20,000 - 30,000	20,000 - 30,000	24,000 - 36,000
GE HL	32	3100	82	3000	80	2.95	25,000	36,000	40,000	45,000
GE SXL	32	2850	81+	2750	80	2.95	31,000	40,000	55,000	60,000
PHILIPS ADV	32	3100	85	3000	82	1.7	24,000	30,000	30,000	36,000
PHILIPS PLUS	32	2950	85	2850	82	1.7	30,000	36,000	38,000	44,000
PHILIPS ADV XLL	32	2950	85	2850	82	1.7	40,000	46,000	46,000	52,000
SYLVANIA XP	32	3000	85	3000	85	3.5	24,000	40,000	40,000	42,000
SYLVANIA XPS	32	3100	85	3100	81	2.9	24,000	40,000	40,000	42,000
SYLVANIA XP/XL	32	2950	85	2950	81	3.5	36,000	52,000	65,000	67,000
GE SPX 28W	28	2725	82	2625	80	2.95	24,000	30,000	45,000	50,000
PHILIPS ADV 28W	28	2725	85	2675	82	1.7	32,000	38,000	38,000	44,000
PHILIPS ADV XLL 28W	28	2675	85	2625	82	1.7	40,000	46,000	46,000	52,000
SYLVANIA XP 28W SS	28	2725	85	2725	81	2.9	24,000	40,000	40,000	42,000
SYLVANIA XP XL 28W SS	28	2600	85	2600	81	5	50,000	75,000	80,000	84,000
GE SPX 25W	25	2400	85	2350	80	2.95	36,000	40,000	50,000	55,000
PHILIPS ADV 25W	25	2500	85	2400	85	1.7	32,000	38,000	38,000	44,000
PHILIPS ADV XLL 25W	25	2400	85	2350	82	1.7	40,000	46,000	46,000	52,000
SYLVANIA XP 25W SS	25	2500	85	2500	81	2.9	24,000	40,000	40,000	42,000
SYLVANIA XP XL 25W SS	25	2400	85	N/A	N/A	5	50,000	75,000	80,000	84,000
F28T5	25-28	2900+	85	2750+	85	1.4 - 2.5	*	*	25,000 - 30,000	30,000 - 40,000
F54T5HO	49-54	5000	85	4800+	85	1.4 - 2.5	*	*	25,000 - 30,000	30,000 - 60,000

Lamp manufacturers may alter rated lamp life and lumen specifications, so get updates from manufacturers.

Prepared by Stan Walercyk of Lighing Wizards 1/1/14 version   www.lightingwizards.com

Table 16-3

CLEAR PRISMATIC LENSED TROFFER - 2014

| $0.15 Blended KWH rate | 1.10 additional air conditioning savings (1.00 is none) | | $0.03 /KWH saved first year rebate for fluorescent | $0.08 /KWH saved first year rebate for LED | $0.06 /KWH saved first year benefit | 15 years of long term benefit |

application and fixture type	watts (existing)	annual hours	annual electric cost	option	retrofit option description	rated lamp life hours @ 3 hour cycles	watts (some averaged)	watts reduction	% watts reduction	annual unit electric cost savings	appr installed cost	rebate	per year maintenance savings benefit for comprehensive long term benefit and payback	per year improved worker productivity from improved lighting quality benefit for comprehensive long term benefit and payback	per year combined maintenance savings and improved worker productivity benefit for long term benefit and payback	payback in years just electricity	payback in years electricity & maintenance	payback in years comprehensive	long term benefit just electricity	Long term benefit electricity & maintenance	long term benefit comprehensive
2x4 clear prismatic lensed troffer with 3 32W 735 20,000 hour rated F32T8s and generic 88 BF ballast	90	3000	$40.50																		
				A	3 25W F32T8 841 or 850 lamps & 71 BF extra efficient program start parallel wired ballast	30,000 - 36,000	56	34	38%	$17	$50	$3	$2	$2	$4	2.8	2.7	2.3	$206	$232	$262
				B	2-cove white reflector, 2 high lumen 32W F32T8 850 lamps & 71 BF extra efficient program start parallel wired ballast	30,000 - 36,000	47	43	48%	$21	$60	$4	$4	$2	$6	2.6	2.4	2.1	$263	$319	$349
				C	high performance hard-wired LED troffer kit with no dimming or Kelvin changing	100,000	35	55	61%	$27	$160	$13	$9	$4	$13	5.4	5.1	3.6	$262	$383	$443
				D	high performance hard-wired LED troffer kit with personally controlled dimming and Kelvin changing (max 35W)	100,000	26	64	71%	$32	$230	$15	$9	$250	$259	6.8	6.5	0.7	$261	$380	$4,130

copyright of Stan Walerczyk of Lighting Wizards, www.lightingwizards.com, 1/1/14 version

Internal Driver

This is the most common type. The existing fluorescent ballast(s) is disconnected from power or totally removed. Line voltage power is wired directly to lamp holders. Here are special concerns with this type, which are heavier because of the driver:

- Some manufacturers may not warranty that their lamp holders will hold these heavier LED T8s.

- Some manufacturers may not warranty their lamp holders for continuous 277V or higher voltage.

- UL has found a fire hazard with some internally shunted lamp holders, which many of the recently new fluorescent T8 fixtures have.

- Some contractors and end users, in order to save time, may use some of the existing wiring from lamp holders and attach that to line voltage wires with wire nuts, which can be an electric code violation because of wrong color of wire and metal gauge.

- Down the road, if an LED T8 lamp is replaced with a fluorescent T8 lamp, and if the line voltage is 277V or higher, there could be significant damage to the glass fluorescent lamps.

Remote Driver

These can be better than ones with internal drivers because these are lighter, and the line voltage wires go to the driver, like they used to go to the ballast. But, they still use existing lamp holders, which is not recommended.

With Existing Instant Start Electronic T8 Ballast

Some electronic ballasts are more efficient than others. How old are they? Maybe they will have to be replaced soon. There is a version of this with at least one manufacturer that sells new high performance instant start ballasts with LED T8s, but these LED T8s still use existing lamp holders.

Are LED T8s more environmentally friendly than fluorescent T8s? Yes; fluorescent lamps have mercury, and LEDs do not. But there are several issues with LEDs, including toxic chemicals used in production, amount of water required in production, and the amount of metal that is mined, melted, and transported for heat sinking. The DOE has not yet determined if LEDs are really more environmentally friendly cradle to cradle, but some LED T8 marketing literature and sales people state "toxic" when referring to fluorescent. Fluorescent lamps now have much less mercury than in the past. Now most fluorescent T8s have 1.7-5.0 mg, considerably less than what is in a typical can of tuna fish. Plus, many areas require fluorescent lamp recycling.

I have also seen LED T8 literature and heard sales people state that fluorescent T12 and T8 lamps only last 10,000-15,000 hours, which is incorrect. However, there are fluorescent T8s rated for up to 67,000 hours, which is longer than the ratings on most LED T8s.

At least through early 2014, DesignLights Consortium (DLC), for the most part, did not really approve of LED T8s in its LED T8 lamp category, but it did approve some in its "lamp-style" retrofit kits for the linear panels category. Again, two or more typical LED T8s can qualify in this category, mainly because of the higher lumens from more than one lamp in this pseudo "kit." Is this gamesmanship?

It is my understanding that the California large investor-owned utilities (including Pacific Gas & Electric, San Diego Gas & Electric, and Southern California Edison), which generally provide rebates on DLC listed products, have not rebated T8s (at least through 2013), even if they are approved one way or the other by the DLC. Hopefully, other rebate organizations will also not rebate LED T8s.

In general, it is usually not a good idea to cram LED technology into incumbent shapes, including 1"-diameter and 4'-long lamps. I have asked manufacturers to stop trying to sell LED T8s in North America and to develop LED light bars, which could be about 1" wide and 4' long. These could be screwed into the top of the fixture, using the entire fixture as a heat sink. Having the light higher in the fixture could also help with improving side light out of the fixture. The driver could be installed in the ballast compartment, which would be better in regard to heat for both the LEDs and the driver. The existing lamp holders would not be used.

There are already manufacturers of good LED light bars, including:

- Albeo (purchased by GE)
- Cree
- LED Living Technology

PlanLED and other manufacturers are developing models. Since LED light bars cost about the same as LED T8s but have several advantages without any disadvantages, why even consider LED T8s?

Usually, hardwired LED lightbars or troffer kits with an integral optical compartment and lens are better than LED light bars. That is discussed in detail in Chapter 17. Like many others, that chapter also has a reference to human centric lighting, because there are some troffers and troffer kits with dimming and Kelvin-changing LEDs.

Chapter 17

LED Lightbar Kits, Troffer Kits and Troffers

INTRODUCTION

For LED light bars, troffer kits and troffers, and many other fixtures to be cost effective compared to high performance incumbent technologies, the LED products usually have to provide at least 100 LPW out of the fixture and have a ten-year warranty. Some manufacturers list 100,000 hours and even considerably longer, but they only have a five-year warranty. I do not care that much about rated lives; I just want to know how long the manufacturers will stand by behind their product. Since there are enough manufacturers who will offer ten-year warranties, that is usually what I require. Also, end customers are often hesitant to approve energy efficient lighting projects when the warranty expires before the payback period, and since paybacks are often longer than five years with LED products, the warranty on those products should be ten years.

LED lightbars, troffer kits and troffers started to give even the best fluorescent retrofits a run for the money in 2013, and by the end of 2014 the LED kits will probably surpass fluorescent. Even if the hardwired LED troffer kits cost significantly more than fluorescent options, at this writing some of the LED products can save 15-35% more energy, have 100,000+ hour rated life, and have up to a ten-year warranty. Plus, the rebates can often be higher on LED products.

LED cost will continue to decrease, and electrical savings will continue to increase. Plus, dimming LED is much better than dimming fluorescent.

Unless the economy really improves, retrofitting existing troffers with hardwired troffer kits or fluorescent lamps and ballasts (and maybe also reflector or upscale kits) is much more significant than new LED or fluorescent troffers in remodels and new construction. Yes, new LED or fluorescent troffers could replace existing ones, but it is often much less expensive to retrofit existing troffers. Union and prevailing-wage projects usually require much higher hourly labor rates for installing new fixtures, sometimes more than double than for retrofitting. Also, retrofitting can be easier and faster; some kits can be installed in five to ten minutes. Plus, you do not need to worry about asbestos or other dust when removing troffers.

Most of what applies to hardwired LED troffer kits also applies to LED troffers. If you have to get new troffers, I would not even consider fluorescent for most projects. Many end customers do not have the money or do not want to spend the extra money on hardwired LED troffer kits or LED troffers, especially if they simply make decisions on hard savings and payback (which are insufficient reasons).

Warm to cool white Kelvin or correlated color temperature (CCT) changing, which can provide substantial human centric lighting benefits, is forecast to be the main reason why LED will win the race against fluorescent in troffers and other fixture types in retrofits, remodels, and new construction in both residential and commercial facilities.

LED T8s are not recommended in existing or new troffers because of performance, pricing, and mainly safety issues. Chapter 16 further discusses LED T8s.

FLUORESCENT

High lumen and full wattage T8 lamps and high performance electronic ballasts can be considered the highest performance fluorescent system, providing about 100 initial LPW, which is quite good.

In Table 17-1, the two far right columns are the most important ones. Table 17-2 shows that you can have the most lumens or the longest life, but not both from the same lamp.

Table 17-1

4' LINEAR FLUORESCENT EFFICACY TABLE

A – 4' lamp type	B – initial catalog or photopic lamp lumens	C – lamp watts	D – lamp lumens per lamp watts	E – lamp quant	F – ballast type	G – standard ballast factor	H – system watts	I – initial system lumens	J – initial system lumens per watt	K – mean or 8000 hour lumen maintenance	L – mean or 8000 hour system lumens	M – mean or 8000 hour system lumens per watt	N – percent light compared to best
high performance F32T8	3100	32	96.9	2	HP IS	0.87	53	5394	101.8	95%	5124	96.7	100%
	3100	32	96.9	2	HP PS	1.15	70	7130	101.9	95%	6774	96.8	100%
	3100	32	96.9	2	G IS	0.87	58	5394	93.0	95%	5124	88.4	91%
extra long life 2950 lumen F32T8	2950	32	92.2	2	HP IS	0.87	53	5133	96.8	95%	4876	92.0	95%
	2950	32	92.2	2	G IS	0.87	58	5133	88.5	95%	4876	84.1	87%
basic grade F32T8	2800	32	87.5	2	HP IS	0.87	53	4872	91.9	95%	4628	87.3	90%
	2800	32	87.5	2	G IS	0.87	58	4872	84.0	95%	4628	79.8	82%
30W F32T8	2850	30	95.0	2	HP IS	0.87	51	4959	97.2	95%	4711	92.4	95%
	2850	30	95.0	2	G IS	0.87	55	4959	90.2	95%	4711	85.7	89%
28W F32T8	2750	28	98.2	2	HP IS	0.87	48	4785	99.7	95%	4546	94.7	98%
	2750	28	98.2	2	G IS	0.87	51	4785	93.8	95%	4546	89.1	92%
25W F32T8	2440	25	97.6	2	HP IS	0.87	42	4246	101.1	95%	4033	96.0	99%
	2440	25	97.6	2	G IS	0.87	47	4246	90.3	95%	4033	85.8	89%
extra long life 25W F32T8	2400	25	96.0	2	HP IS	0.87	42	4176	99.4	95%	3967	94.5	98%
	2400	25	96.0	2	G IS	0.87	47	4176	88.9	95%	3967	84.4	87%
high lumen F28T5	3050	28	108.9	2	HP PS	0.95	58	5795	99.9	93%	5389	92.9	96%
typical F28T5	2900	28	103.6	2	G PS	1.00	64	5800	90.6	93%	5394	84.3	87%
26W F28T5	2900	26	111.5	2	HP PS	0.95	55	5510	100.2	92%	5069	92.2	95%
49W F54T5HO	5000	49	102.0	2	HP PS	1.00	105	10000	95.2	93%	9300	88.6	92%
typical F54T5HO	5000	54	92.6	2	G PS	1.00	117	10000	85.5	93%	9300	79.5	82%
F34T12 800	3100	34	91.2	2	G RS E	0.85	60	5270	87.8	93%	4901	81.7	84%
F34T12 CW	2650	34	77.9	2	G RS M	0.88	72	4664	64.8	87%	4058	56.4	58%

notes: Lumens, lumen maintenance, ballast factors and wattages may vary among various manufacturers.
In enclosed fixtures, since reduced wattage F32T8s consume less heat they can often operate closer to optimal 77 degrees F temperature, so may provide more light than this table shows compared to full wattage.
Although efficacy can be improved with IS and RS ballasts with T5s and T5HOs, lamp life can be greatly reduced and lamp manufacturers may not warranty lamps.
93% is used as an average EOL lumen maintenance for T5HOs. 90% - 94% range among manufacturers.
All wattages based on 277V. HP is high performance. G is generic. IS is instant start. PS is program start. RS is rapid start. E is electronic. M is magnetic.
Extra long life is 36,000 hours with IS and 40,000 hours with PS ballasts at 3 hour cycles.
Prepared by Stan Walerczyk of Lighting Wizards www.lightingwizards.com 1/1/14 version

Table 17-2

4' T8 LAMP LIFE, LUMENS, CRI & MERCURY

LAMP	WATTS	3000-4100K		5000K		MAX MG OF HG	LAMP LIFE HOURS			
							INSTANT START		PROGRAM START	
		LUMENS	CRI	LUMENS	CRI		3 HR	12 HR	3 HR	12 HR
1st GENERATION - GENERIC	32	2800	75-78	2800	75-78	1.7 - <10	15,000 - 24,000	20,000 - 30,000	20,000 - 30,000	24,000 - 36,000
2nd GENERATION - GENERIC	32	2950	81-85	2800 - 2950	80-85	1.7 - <10	15,000 - 24,000	20,000 - 30,000	20,000 - 30,000	24,000 - 36,000
GE HL	32	3100	82	3000	80	2.95	25,000	36,000	40,000	45,000
GE SXL	32	2850	81+	2750	80	2.95	31,000	40,000	55,000	60,000
PHILIPS ADV	32	3100	85	3000	82	1.7	24,000	30,000	30,000	36,000
PHILIPS PLUS	32	2950	85	2850	82	1.7	30,000	36,000	38,000	44,000
PHILIPS ADV XLL	32	2950	85	2850	82	1.7	40,000	46,000	46,000	52,000
SYLVANIA XP	32	3000	85	3000	85	3.5	24,000	40,000	40,000	42,000
SYLVANIA XPS	32	3100	85	3100	81	2.9	24,000	40,000	40,000	42,000
SYLVANIA XP/XL	32	2950	85	2950	81	3.5	36,000	52,000	65,000	67,000
GE SPX 28W	28	2725	82	2625	80	2.95	24,000	30,000	45,000	50,000
PHILIPS ADV 28W	28	2725	85	2675	82	1.7	32,000	38,000	38,000	44,000
PHILIPS ADV XLL 28W	28	2675	85	2625	82	1.7	40,000	46,000	46,000	52,000
SYLVANIA XP 28W SS	28	2725	85	2725	81	2.9	24,000	40,000	40,000	42,000
SYLVANIA XP XL 28W SS	28	2600	85	2600	81	5	50,000	75,000	80,000	84,000
GE SPX 25W	25	2400	85	2350	80	2.95	36,000	40,000	50,000	55,000
PHILIPS ADV 25W	25	2500	85	2400	85	1.7	32,000	38,000	38,000	44,000
PHILIPS ADV XLL 25W	25	2400	85	2350	82	1.7	40,000	46,000	46,000	52,000
SYLVANIA XP 25W SS	25	2500	85	2500	81	2.9	24,000	40,000	40,000	42,000
SYLVANIA XP XL 25W SS	25	2400	85	N/A	N/A	5	50,000	75,000	80,000	84,000
F28T5	25-28	2900+	85	2750+	85	1.4 - 2.5	*	*	25,000 - 30,000	30,000 - 40,000
F54T5HO	49-54	5000	85	4800+	85	1.4 - 2.5	*	*	25,000 - 30,000	30,000 - 60,000

Lamp manufacturers may alter rated lamp life and lumen specifications, so get updates from manufacturers.

Prepared by Stan Walercyk of Lighing Wizards 1/1/14 version www.lightingwizards.com

If you put them into a lensed or parabolic louvered troffer, the LPW out of the fixture can range from the 50s to the 80s, and LPW out of the fixture is much better than just fixture efficiency.

There are several culprits. Light rays that hit a lens too low (with any angle) are reflected back into the fixture. Most of the time a light ray is reflected from even a highly reflective surface; the loss may be 5-10%, and light rays may bounce around several times before getting out the fixture. There is also light loss from light rays bouncing between adjacent lamps. Numerous fluorescent troffers have vertical or nearly vertical sides, so much of the light just goes side to side and not out of the fixture. Full wattage T8s have maximum light output at 77°F, and reduced wattage T8s, all T5s, and all T5HOs have maximum light output at 90-95°F. Many lensed and non-air-handling troffers with 2-4 lamps can get well over 100°F after an hour or so, which can reduce light from the lamps 10-25%. Fixture efficiency does not really include thermal factors. Linear fluorescent lamps direct most of their light perpendicular to lamp length and very little out from the ends, so distribution is not equal in all directions.

High performance, full wattage T8 lamps can cost $2.50-$5.00 and are rated up to 67,000 hours, with up to a five-year parts and labor warranty. High performance, fixed output electronic ballasts can cost $10-$25, are rated for 60,000 hours, and typically have a 5-year parts and labor warranty. A 4' reflector kit can cost $10-$15. Retrofitting with lamp-for-lamp and new ballasting may cost $40-$50 with a 2-4 lamp 2x4 troffer, including parts and labor, for the end customer. Retrofitting a 3-4 lamp 2x4 troffer with a reflector kit, 2 lamps, and a new ballast may cost $60-$65, including parts and labor.

Although several manufacturers and sales companies push dimming fluorescent ballasts, even the best ones are not very efficient when dimmed below .70 BF. That is when the ballasts need to heat the lamp cathodes to prevent the lamps from spiraling, flickering, or turning off. This cathode heating consumes substantial wattage. For example, dimming every dimming fluorescent ballast to 50% light level can consume 20-30% more wattage than turning off every other equivalent fixed BF ballast.

Some of the new high performance troffers and retrofit kits are designed for one high lumen T8 lamp, and these can have quite good lumens per watt out of the fixture. Fluorescent troffers, now that most lenses are UV stabilized, can last practically forever with lamp and ballast replacements.

Yes, fluorescent lamps have mercury, but mercury levels have been significantly reduced, such as to 1.75 mg in some 4' T8s. Some cans of tuna fish have more mercury than two T8s. Plus, many areas require recycling.

Although most providers are still rebating fluorescent projects, some, including Pacific Gas & Electric, San Diego Gas & Electric and Southern California Edison (the big three investor-owned electric and natural gas utilities in California) have reduced the rebate on fluorescents from $.05/kWh saved over the first year to .03/kWh and in 2013 increased rebates on LED to $.08/kWh. Other rebate providers across North America may be doing something similar.

With the evolution of LED and other solid state lighting (SSL), that is where most manufacturers are investing their research. Very little development is going into mature technologies, including fluorescent. It has already started to become an LED world.

Clear Prismatic Lensed Fluorescent Troffers

As you probably know, these are the most common troffers in North America because they are relatively inexpensive, with decent performance. (See Figure 17-1.)

One positive attribute is that there is a fairly good downward batwing distribution, especially perpendicular to lamp length, allowing fairly even light levels between and underneath troffers. (In Figure 17-2, the solid line is perpendicular to lamp length.) These can be easily and inexpensively retrofitted lamp-for-lamp, or delamped with or without a reflector kit.

Figure 17-1. Clear Prismatic Lensed Fluorescent Troffer

Parabolic Fluorescent Troffers

These were very popular in the late 80s and early 90s because they reduced glare on the curved glass video display terminals (VDTs). However, now most computer screens are flat, with much better resolution and anti-glare coatings, so the disadvantages of parabolic and paracube louvers are much more apparent. Following are some of the disadvantages in the millions and millions of these troffers across North America and the rest of the world.

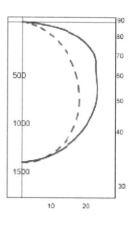

Figure 17-2. Diagram of Clear Prismatic Lensed Fluorescent Troffer

- 70-75% fixture efficiency
  - 25-30% of the light from lamps never gets out of the fixture.
  - Including thermal effect, less than 70-75 LPW gets out of the fixture, even with the best fluorescent lamps and ballasts.

- Dreaded cave effect
  - Dark ceilings and upper walls make the space seem small and gloomy.

- Insufficient vertical foot-candles
  - It can be difficult to read calendars and other items on walls.

- High contrast ratios
  — With most of the light coming straight down, with very little going sideways, and being angled down to the side, contrast ratios can be excessive, which can cause eye strain and headaches.

- Overhead glare
  — With no lens between the fixture and eyes, even when looking horizontally underneath one of the fixtures, the light can cause eye strain and headaches.

- Ultraviolet (UV) exposure
  — With no lens, people can be exposed to below 460 nanometers of light, which is UV and can cause macular degeneration and other eye problems.

Figure 17-3 shows examples of why people often do not like parabolic troffers. The scene on the left has pillow cases attached to the T-bar. The one on the right goes an additional step, with fabric across the cubicle walls.

Figure 17-4 is my most infamous example of the dislike for parabolic troffers. The facility manager told me that this office worker also put styrofoam around the inside of the tent to reduce the noise. He said it took him two months for this worker to remove this tent. (He also said that this worker should work out of the house.)

Several manufacturers provide upscale fluorescent kits that eliminate the outdated parabolic louvers, improve lighting quality, often reduce the number of lamps from three to one, and often reduce wattage by two thirds to three quarters. Even if the parts cost, including lamping and ballast, may be about $65, and the total installed price may be $110-$120, the financial returns are usually quite good.

The following companies have one- and two-lamp kits that totally eliminate parabolic louvers and are highly recommended in offices, classrooms, and halls.

Figure 17-3. Problematic parabolic troffers

Figure 17-4. One worker's extensive coping strategy for parabolic troffers

- A.L.P.
- Energy Solutions International
- Envirobrite-Energy Planning Associates
- Daybrite
- Lithonia

Figure 17-5 shows a good example.

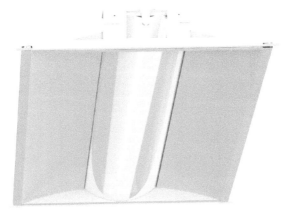

Figure 17-5. Envirobrite Dailite lamp kit

There are some products, including Lithonia's ES8R high performance parabolic kit, which can work well in some retail stores. You may be able to see these in some Sears and Target locations.

One big problem is that too many lighting retrofit contractors and ESCOs retrofit parabolic troffers lamp-for-lamp, or delamp with reflectors and repositioned lamps, which makes the cut-off angles even worse and often creates a glare bomb. Several end customers had me re-retrofit delamped parabolic troffers because of too much glare.

Please help get rid of every parabolic louver, as well as every paracube louver, its cousin. This can be done cost effectively with upscale fluorescent kits and hardwired LED kits that have integral lenses.

LED

While dimming ballasts get less efficient when dimming below .70 BF, LED systems can maintain or become more efficient when dimmed because they are more efficient when cooler—and they get cooler when dimmed. It is important to get dimming LED systems with little or no flicker when dimming.

The diagram in Figure 17-6 shows how LED chips are improving and getting less expensive decade after decade, which also translates to year after year. With higher lumen LEDs, fewer are needed, which helps bring down the cost. Also, production cost for LEDs is coming down.

Hardwired LED Light Bar Troffer Kits

There are two types of hardwired LED troffer kits, and both should have easily replaceable LEDs and drivers. The kits are often about one inch wide and two or four feet long. They can be screwed into the top of the fixture, using the metal frame as a heat sink. Since the LEDs are higher than the LEDs in LED T8s, there can be more side light. Plus, these kits do not use existing lamp

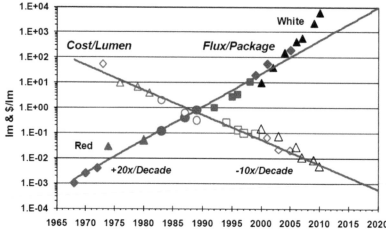

Figure 2.4: Haitz's Law: LED Light Output Increasing / Cost Decreasing
Source: Roland Haitz 2010

Figure 17-6. LED light output and cost

holders, which can be a big safety benefit. The driver can go into the ballast compartment, which separates heat from the driver and LEDs.

Two of these bars with a remote driver may cost $100 or less in the future. This is about the same as for two LED T8s. Although these bars can be significantly better than LED T8s, they still keep the existing lenses or louvers, so the amount of light and distribution can vary greatly in various troffers. These kits may be better in strip, hooded industrial, and seal or vapor tight fixtures than troffers.

Some organizations did not provide rebates on these types of kits until early 2014, but hopefully that will change. Several manufacturers provide these kits as of early 2014, including:

- Cree
- LED Living Technology
- RedBird

By early 2014, RedBird's Cardinal StripItKits have 130 LPW out of a lensed fixture, 100,000 hour rated life and ten year warranty. Numerous other manufacturers will probably develop LED lightbar kits. LED lightbars may be very good for strips, hooded industrials and wraps.

Hardwired LED Troffer Kits with Integral Optics

In 2012, LED troffers and LED hardwired troffer kits, with integral optics and about 90 LPW out of the troffer, started to compete fairly well with the best fluorescent troffers and retrofit solutions. In that year, there were some LED troffers and troffer kits with 100 or 110 LPW, but that was achieved with lower drive current and additional LEDs to provide sufficient light. In most appli-

Figure 17-7. Cree UR Series

cations, the extra wattage savings did not cover the higher cost of products with extra LEDs. By 2014, some manufacturers are providing:

- Up to 120 LPW out of the troffer without under-driving LEDs and needing extra LEDs

- 100,000+ hour rated life

- Up to 10-year parts warranty

- 15-35% energy savings compared to the best fluorescent systems without dimming

- Considerably more energy savings when dimming

These kits cost more than the bar, strip, or pad LED troffer kits, but they are much better. Kits with integral optics that include their own lenses provide the same amount of light and distribution, no matter in which troffers they are installed. Plus, keeping the LEDs and drivers cooler may be easier and more consistent.

Whereas most clear prismatic lensed fluorescent troffers have a good downward batwing distribution in at least two directions, many of the hardwired LED troffer kits with integral optics have a cosine or squashed cosine distribution, so most of the light goes straight down, and it can be dark between fixtures on desks and on the floor. But there are some good ones that have a pretty good downward distribution. You can easily see this when you inspect the polar curves of various products from various manufacturers. Some manufacturers of hardwired LED troffer kits in 2013 that have integral lenses are:

- Cooper
- Cree
- Energy Solutions International
- Envirobrite
- Harris Lighting
- Lithonia

- Osram Sylvania
- Philips
- PlanLED

For example, here is some information on the 2x2 and 2x4 Envirobrite and PlanLED troffer kits, which can fit into most standard and shallow lensed parabolic and paracube troffers:

- 100 + LPW

- Up to 100 LPW in mid-2013

- 100,000+ hour rating at L70

- 10-year warranty when operated no more than 4380 annual hours
  — 10-year warranty may be available for 8760 annual hours, if slightly more is paid initially

- 80+ color rendering index (CRI) with positive R9 for reds

- Better downward batwing distribution than from many other manufacturers

- Non-dimming and fixed Kelvin, dimming and fixed Kelvin, and dimming and Kelvin-changing options
  — Low flicker when dimming

- $100-$200 pricing, depending on quantity and features
  — Controls are extra

A sample kit is shown in Figure 17-8.

Figure 17-8. Envirobrite PlanLED kit

SPECTRALLY ENHANCED LIGHTING

Although Chapter 6 is dedicated to this, as a reminder, high CCT fluorescent or LED is recommended in most applications because of wattage savings while maintaining or improving visual acuity without increasing wattage. Spectrally enhanced lighting can also be used with LED, and that is one reason that I often want at least 5000K from LED products. At the same Kelvin, fluorescents and LEDs have different spectral distribution curves; it will take some time to measure the scotopic/photopic (S/P) ratios of LEDs from various manufacturers.

HUMAN CENTRIC LIGHTING

There is also a separate chapter on this (29), which discusses how important it can be to have different light levels and Kelvin at different times of the day for different tasks. A website that can be very helpful is: http://humancentriclighting.com

Some of the tunable LED products have the same lumens, wattage, LPW, life, warranty, and cost as standard LED products. The added cost of Kelvin-changing on top of dimming and other controls is miniscule. If you are planning to dim, why not also get tunable LED products?

FINANCIALS

As mentioned in Chapter 24 (Financial Tools), payback is not a good financial tool, especially for long life LED products, because payback does not include any benefit after the payback period.

SOFT SAVINGS

This is also discussed in Chapter 24, which describes in detail how soft savings (including improved worker productivity, student test scores, and retail sales) usually dwarf hard savings (electrical savings, rebates, and sometimes also maintenance savings).

Although soft savings can be difficult to quantify, that can be accomplished. Soft savings often need to be seen and experienced to be believed. You could do a mock-up, especially with tunable LED products, and at least anecdotally evaluate how people like them. Human resource offices are often a very good place for this, because those people understand the value of worker satisfaction and productivity.

LED & FLUORESCENT RETROFIT COMPARISONS

Tables 17-3 through 17-4 display comparisons for lensed and parabolic troffers. Your projects may have different kWh rates, air conditioning savings, rebates for fluorescent and LED, and annual hours of operation, which you could use in your own tables. Proposed wattages for each option should be modified if existing light levels are on the low or high side, or the end customer wants higher or lower light levels for certain reasons. Installed costs may be higher or lower depending on the size of the project, location, and whether union, prevailing wage, or federal wage structures are required.

The shaded contiguous columns in these tables are subjective, and the values could easily be increased or decreased. Regarding maintenance savings, sometimes just avoided parts savings can be used, and sometimes avoided labor savings can also be used. Either way, there are more savings when there is delamping and going with long life LED than with lamp-for-lamp. Regarding worker productivity, as you can see, there is not very much improvement with clear prismatic lensed troffers unless personally controlled tunable LEDs are included. For the parabolic troffers, anything that gets rid of the parabolic louvers

Table 17-3

CLEAR PRISMATIC LENSED TROFFER - 2014

$0.15 | blended KWH rate | 1.10 — additional air conditioning savings (1.00 is none) — $0.03 /KWH saved first year rebate for fluorescent | $0.08 /KWH saved first year rebate for LED — copyright of Stan Walerczyk of Lighting Wizards, www.lightingwizards.com, 1/1/14 version

		existing						proposed												15 — years of long term benefit		
application and fixture type	watts	annual hours	annual electric cost	option	retrofit option description	rated lamp life hours @ 3 hour cycles	watts (some averaged)	watts reduction	% watts reduction	annual unit electric cost savings	appr installed cost	rebate	per year maintenance savings benefit for comprehensive long term benefit and payback	per year improved worker productivity from improved lighting quality benefit for comprehensive long term benefit and payback	per year combined maintenance savings and improved worker productivity benefit for long term benefit and payback	payback in years just electricity	payback in years electricity & maintenance	payback in years comprehensive	long term benefit just electricity	Long term benefit electricity & maintenance	long term benefit comprehensive
2x4 clear lensed prismatic troffer with 3 32W 735 20,000 hour rated F32T8s and generic 88 BF ballast	90	3000	$40.50	A	3 25W F32T8 841 or 850 lamps & .71 BF extra efficient program start parallel wired ballast	30,000 - 36,000	56	34	38%	$17	$50	$3	$2	$2	$4	2.8	2.7	2.3	$206	$232	$262
				B	2-cove white reflector, 2 high lumen 32W F32T8 850 lamps & .71 BF extra efficient program start parallel wired ballast	30,000 - 36,000	47	43	48%	$21	$60	$4	$4	$2	$6	2.6	2.4	2.1	$263	$319	$349
				C	high performance hard-wired LED troffer kit with no dimming or Kelvin changing	100,000	35	55	61%	$27	$160	$13	$9	$4	$13	5.4	5.1	3.6	$262	$383	$443
				D	high performance hard-wired LED troffer kit with personally controlled dimming and Kelvin changing (max 35W)	100,000	26	64	71%	$32	$230	$15	$9	$250	$259	6.8	6.5	0.7	$261	$380	$4,130

Table 17-4

PARABOLIC TROFFER - 2014

$0.15 | blended KWH rate | 1.10 — additional air conditioning savings (1.00 is none) — $0.03 /KWH saved first year rebate for fluorescent | $0.08 /KWH saved first year rebate for LED — copyright of Stan Walerczyk of Lighting Wizards, www.lightingwizards.com, 1/1/14 version

		existing						proposed												15 — years of long term benefit		
application and fixture type	watts	annual hours	annual electric cost	option	retrofit option description	rated lamp life hours @ 3 hour cycles	watts (some averaged)	watts reduction	% watts reduction	annual unit electric cost savings	appr installed cost	rebate	per year maintenance savings benefit for comprehensive long term benefit and payback	per year improved worker productivity from improved lighting quality benefit for comprehensive long term benefit and payback	per year combined maintenance savings and improved worker productivity benefit for long term benefit and payback	payback in years just electricity	payback in years electricity & maintenance	payback in years comprehensive	long term benefit just electricity	Long term benefit electricity & maintenance	long term benefit comprehensive
2x4 18 cell parabolic troffer with 3 32W 735 20,000 hour rated F32T8s and generic 88 BF ballast	90	3000	$40.50	A	3 25W F32T8 841 or 850 lamps & .71 BF extra efficient program start parallel wired ballast	30,000 - 36,000	56	34	38%	$17	$50	$3	$2	$2	$4	2.8	2.7	2.3	$206	$232	$262
				B	upscale kit which eliminates parabolic louvers, 1 high lumen F32T8 841 or 850 lamp & 1.15 BF extra efficient program start ballast	30,000 - 36,000	39	51	57%	$25	$115	$5	$6	$250	$256	4.4	4.1	0.4	$268	$354	$4,104
				C	high performance hard-wired LED troffer kit with no dimming or Kelvin changing	100,000	28	62	69%	$31	$160	$15	$9	$250	$259	4.7	4.4	0.5	$315	$435	$4,185
				D	high performance hard-wired LED troffer kit with personally controlled dimming and Kelvin changing (max 28W)	100,000	18	72	80%	$36	$230	$17	$9	$500	$509	6.0	5.7	0.4	$322	$440	$7,940

is a big improvement, and the biggest improvement is the personally controlled tunable LED system.

Payback of 15 years is used for long-term benefit because that is typical fluorescent ballast life. If you know that a lease will expire in five years, or that the building will be demolished in 10 or so years, the number of years could be changed.

In each table, there are six financial columns on the right side, and the bolded comprehensive long-term benefit is considered the best, if the customer will accept it.

The tables show that the lamp-for-lamp retrofit with 25W F32T8s has the lowest installed cost and the best payback with just electricity, but the lowest wattage and maintenance savings. Plus, it hardly does anything to improve lighting quality other than increasing CRI from 75-78 up to 80-85.

The tables show that the LED solutions cost significantly more initially and have worse paybacks with just electricity, but they save the most and have quite good comprehensive financial returns.

As you can see in Table 17-3, options B and C have almost identical long-term comprehensive benefits. Even though option C costs three times more initially, many end customers may want it because of lower maintenance costs, the wow factor of LED, or other reasons. Option D, which costs the most, provides, by far, the greatest long-term comprehensive benefits.

A term of 15 years is used for long-term benefit because that is typical ballast life, when half are expected to be burned out and half are expected to still be working.

You can also see that going lamp-for-lamp with reduced wattage fluorescent T8s is not very good, even though many ESCOs and lighting retrofit contractors do that.

How does 2014 look for a clear prismatic lensed troffer? Well, LEDs look better in comparison.

How does 2014 look for a parabolic troffer? Again, LEDs look better in comparison.

TASK AMBIENT LIGHTING

See Chapter 18 on this, which is very relevant to this topic. In Tables 17-3 through 17-4, if tunable LED task lights were included, the troffer wattage and ambient light levels could be reduced.

FUNDING

Even though hardwired LED troffer kits, especially tunable ones, have much better long-term benefits, it can often be a challenge to pay the higher initial cost.

Positive cash flow financing is one example of funding. Every month the savings are greater than what has to be paid to the manufacturer, finance company, bank, or ESCO. When the project is paid for, the end customer gets all of the savings.

At this writing, the large, California investor-owned utilities have on-bill financing at zero or low interest. A qualified customer signs up and gets the project installed. The monthly electric bills stay the same, with the savings paying off the consulting, parts, and labor costs of the project. When all of that is paid, the customer's bill goes down. The Pacific Gas & Electric website with information on this program is:
www.pge.com/mybusiness/energysavingsrebates/rebatesin-centives/taxcredit/onbillfinancing

Hopefully, other utilities across North America will offer similar financing.

EPACT TAX DEDUCTIONS

The previous U.S. federal government tax deductions expired end of 2013 and hopefully there will a new progam.

WAITING TO RETROFIT

Some people want to wait up to a few years to retrofit until LED products get better and less expensive, but often the lost savings is never recouped. This can be checked for specific projects, using good long-term financials. One expert on this is Mark Jewell at EEFG, Inc. (www.eefg.com)

DESIGNLIGHTS CONSORTIUM (DLC)

Many organizations only rebate commercial LED products if they are DLC approved. Although it is assumed that the DLC has good intentions in trying to protect end customers, it has pushed people away from hardwired troffer kits and troffers and toward fluorescent retrofits and new troffers.

In early 2013, DLC did reduce minimal lumen requirements on hardwired troffers and troffer kits, but this was not sufficient to actually allow these LED products to really compete with fluorescent.

One example where task ambient lighting is the best solution is a typical open office space. With good task lights, the ambient troffers really should only provide 10-20 foot-candles, or 100-200 lux, on desks. That can be easily be achieved by delamping two-, three-, or four-lamp 2x4 troffers down to one 3100-lumen F32T8 lamp, one-cove white reflector, and a .71 to .89 BF high performance ballast, which can provide about 1760-2210 lumens out of the fixture, consume 24 to 29 watts, and usually qualify for significant rebates. But, at mid-2013, the DLC required a minimum of 3000 lumens out of 2x4 LED troffers or troffer kits in order to qualify for rebates. But, in early 2014, the DLC required a minimum of 3000 lumens out of 2x4 LED troffers or troffer kits in order to qualify for rebates. With computer work, often more light is worse than less light, and the 3000 lumens may be too much. Yes, a 3000 lumen LED troffer or troffer kit could be dimmed, but the extra LEDs and dimming system increases the cost.

At least as of this writing, the DLC is also very shortsighted in mandating maximum 5000K in LED troffer kits and troffers, when the human centric lighting benefits from higher-Kelvin lighting at certain times of the day for certain tasks are so apparent. Fluorescent lamps of 6500, 8000, and even higher Kelvin can be used successfully and often qualify for rebates. Why should the DLC or any other organization require minimal lumens and Kelvin, when that should be decided by lighting professionals and end customers? Maybe by late 2013, the DLC will eliminate or adequately reduce minimum lumen requirements and eliminate maximum CCT requirements. If not, you could contact them at: www.designlights.org

The Lighting Design Lab (LDL) has the LDL LED Qualifying Products List, which some rebate organizations use. This list is much better than the DLC Qualified Products List, because the LDL List requires minimum LPW and CRI but not minimum lumens or Kelvin, which is practical. (http://lightingdesignlab.com)

CONCLUSION

2013's Lightfair had the biggest shift toward LED and away from incumbent technologies in the exhibit halls, compared to any past one-year shift.

Hopefully, you understand that although LED has been really making strides and will be the future of lighting, there are still some projects, especially with low kWh rates and low or no rebates, where fluorescent is currently the best choice with certain customers, kWh rates, and rebate amounts. However, there are also many existing projects where LED, often with dimming and Kelvin changing, is already best, if the lighting vendor and end customers get through the payback and hard savings roadblocks. The traditional ESCO payback and hard savings model will probably have to change, especially with the human centric lighting benefits from tunable LED products.

Since early 2013, about half of my clients have told me they still prefer fluorescent in troffers, while the other half have said they want to go with LED because they know it is becoming an LED world and do not want their buildings to be dated in a few years with old technology. Organizations that provide higher rebates for LED compared to incumbent technologies are pushing the shift toward LED.

There is often a six-months to more than a year delay between initial specification and pricing and when the products are actually ordered and delivered, especially with many ESCO projects. Although fluorescent products may appear to be the most cost effective initially, by the time for ordering and delivery has passed, LED products may be the best solution.

Chapter 18

Task Ambient Lighting

INTRODUCTION

Task ambient lighting is relatively low ambient lighting, coupled with task lights when more light is required. This is one of the best strategies for energy efficiency and good lighting. However, it is often not specified at all, and even when it is, often not optimally. The DOE ranks it number one in the top ten underused technologies for saving energy. Since we now have a new generation of cost effective tunable (dimming and CCT or Kelvin-changing) LED task lights, there is really no good excuse to use anything else, except maybe for industrial applications. These tunable LED task lights can also be very good for human centric lighting.

GENERAL DISCUSSION

Task ambient lighting is one of the best strategies in many applications. Troffers, ceiling surface mount, or suspended fixtures provide a relatively low amount of ambient light, but people can use good task lights when they need more light.

There are several reasons why this is so good. Let's focus on offices. With people primarily doing computer work, less light is usually better than more light. This is because computer screens are self illuminated and excess ambient light often causes glare. When people read paper documents or do other tasks that require more lighting, they can use task lights.

Recommended light levels are based on the combination of ambient and task lighting, not just ambient. I often consider open office spaces glorified hallways, with enough ambient light to walk around in and do computer work, and I recommend about 20 foot-candles on desks. Often each ambient light fixture only needs to consume 20-30W to provide these foot-candles. Task lights may consume 8-12W in full mode. Ambient and task lights together can often provide up to 100 foot-candles or more. My LED desk-mount task light can provide 75 foot-candles by itself.

From a single source, light on a target is lost exponentially as distance increases. For example, if the distance is doubled between source and task, there will be one-fourth the foot-candles. So, if 75 foot-candles are needed to read old paper prints or a book, much less wattage will be required from an efficient task light that is two feet above the desk than from an efficient ambient light that is six feet above the desk.

With well-designed new or retrofitted fluorescent or LED ambient lighting and dimming LED task lights, the power density can be .5 or less in watts per square foot. At 3000 annual hours and a $.15/kWh electric rate, that is less than $.23 per square foot per year. For a hundred square foot office space, that is only $23 for a whole year.

Workers often do not have much control of their space in open-office modules. Giving them task lights which they can at least turn on and off can give them some control, which can increase worker satisfaction and productivity. Dimming is even better.

Even though many existing office modules, and even individual desks, already have undercabinet and/or desk-mount task lights, most of these are not very good. Undercabinet linear fluorescent task lights are usually overlit and are glare bombs, especially if the desk is shiny. Some of the ballasts for T8 lamps are magnetic, and even if the ballasts are electronic, they are usually not that efficient. Although many of the desk-mount task lights are CFL, which are fairly good, way too many are still incandescent or halogen.

There are some very good dimming, fixed-Kelvin LED desk-mount and undercabinet task lights. Table 18-1 shows how

Table 18-1.

TYPICAL OFFICE 12' long x 10' wide x 9' high

$0.15 blended KWH rate	1.05 additional air conditioning savings (1.00 is none)	$0.05 /KWH saved first year rebate		10

application and fixture type	existing watts	annual hours	annual electric cost	option	retrofit option description	rated lamp life hours @ 3 hour cycles	watts (some averaged)	watts per square foot	watts reduction	% watts reduction	annual unit electric cost savings	appr installed cost	rebate	per year maintenance savings benefit for comprehensive long term benefit and payback	per year improved worker productivity from improved lighting quality benefit for comprehensive long term benefit and payback	per year combined maintenance savings and improved worker productivity benefit for long term benefit and payback	payback in years just electricity	payback in years electricity & maintenance	payback in years comprehensive	long term benefit just electricity	Long term benefit electricity & maintenance	long term benefit comprehensive
12 x 10 x 9 office area with 2 2x4 18 cell parabolic troffers, each with 3 32W 735 20,000 hour rated F32T8s and generic 88 BF ballasting (1.51 watts per square foot)	180	3000	$81.00	A	Retrofit each troffer with 3 25W F32T8 5000K lamps & .71 BF extra efficient program start parallel wired ballast	30,000-36,000	112	0.93	68	38%	$32	$120	$10	$2	$2	$4	3.4	3.4	3.0	$212	$221	$241
				B	Retrofit each troffer with upscale kit which eliminates parabolic louvers, 1 high lumen F32T8 5000K lamp & 1.15 BF extra efficient program start ballast	30,000-36,000	78	0.65	102	57%	$48	$230	$15	$6	$250	$256	4.5	4.3	0.7	$267	$312	$2,812
				B1	Retrofit each troffer with upscale kit which eliminates parabolic louvers, 1 high lumen F32T8 5000K lamp & .89 BF extra efficient program start ballast. Also include 8W LED task light.	30,000-36,000 for T8s	67	0.56	113	63%	$53	$310	$17	$6	$350	$356	5.5	5.4	0.7	$241	$284	$3,784
				B2	Retrofit each troffer with upscale kit which eliminates parabolic louvers, 1 high lumen F32T8 5000K lamp & .71 BF extra efficient program start ballast. Also include 2 8W LED task lights.	30,000-36,000 for T8s	62	0.52	118	66%	$56	$380	$18	$6	$400	$406	6.5	6.4	0.8	$195	$238	$4,238
				C	Remove both troffers. Install 8' suspended indirect/direct fixture that has 2 high lumen F32T8 5000K lamps & 1.15 BF extra efficient program start ballast.	30,000-36,000	70	0.58	110	61%	$52	$410	$17	$6	$350	$356	7.6	7.5	1.0	$126	$170	$3,670
				C1	Remove both troffers. Install 8' suspended direct/indirect fixture that has 2 high lumen F32T8 5000K lamps & .89 BF extra efficient program start ballast. Also include 8W LED task light.	30,000-36,000 for T8s	63	0.53	117	65%	$55	$490	$18	$6	$450	$456	8.5	8.4	0.9	$80	$123	$4,623
				C2	Remove both troffers. Install 8' suspended direct/indirect fixture that has 2 high lumen F32T8 5000K lamps & .71 BF extra efficient program start ballast. Also include 2 8W LED task lights.	30,000-36,000 for T8s	60	0.50	120	67%	$57	$560	$18	$6	$500	$506	9.6	9.5	1.0	$25	$67	$5,087
				F2	Retrofit each troffer with 3 15W LED T8 lamps. Also include 2 8W LED task lights to provide sufficient light.	25,000-50,000	104	0.87	76	42%	$36	$600	$0	$2	$2	$4	16.7	16.7	15.0	-$241	-$221	-$201
				G	Retrofit each troffer with high performance hardwired LED troffer kit with batwing distribution, set at 35W, so sufficient light at end of life.	50,000+	70	0.58	110	61%	$52	$270	$17	$12	$250	$262	4.9	4.6	0.8	$266	$370	$2,870
				G1	Retrofit each troffer with high performance hardwired LED troffer kit with batwing distribution, set at 30W. Also include 8W LED task light.	50,000+	60	0.50	120	67%	$57	$340	$18	$12	$350	$362	5.7	5.5	0.8	$245	$347	$3,847
				G2	Retrofit each troffer with high performance hardwired LED troffer kit with batwing distribution, set at 25W. Also include 2 8W LED task lights.	50,000+	55	0.46	125	69%	$59	$420	$19	$12	$400	$412	6.8	6.6	0.9	$189	$291	$4,291

copyright of Stan Walerczyk of Lighting Wizards, www.lightingwizards.com, 1/1/14 version

including one or two of these in a private office can help save wattage and improve worker productivity. Since payback does not include any benefit after the payback period, payback is not a good financial tool. The best financial tool in this table is long-term benefit comprehensive, at the far right.

Please be aware that improving worker productivity just 1% is $500 per year, year after year, for someone making $50,000 per year. That 1% can be merely wasting 5 minutes less per 8-hour shift.

PlanLED, Finelite, Koncept, Luxo, Philips, Steelcase, and numerous others offer good fixed-Kelvin, LED desk-mount and undercabinet task lights. Another good fixed-Kelvin option is a furniture-integrated, one-piece system that provides both ambient and task lighting. An example is Tambient.

TUNABLE LED TASK LIGHTS

Tunable desk-mount LED task lights are often the easiest introduction to human centric lighting. They have about the same lumens per watt efficacy and cost about the same as fixed-Kelvin LED task lights, so tunable LED task lights are cost effective just based on energy savings. If the soft savings of improved worker satisfaction and productivity are included, they are pretty much a slam dunk.

People can turn on and off, aim, dim, and adjust color tone with these task lights. Following are some websites for videos on PlanLED Prism models, which show dimming and Kelvin changing. If these addresses do not work in the future, you could probably do a YouTube or web search to find some. www.youtube.com/watch?v=ULc50FSMCPE&feature=em-share_video_user

www.youtube.com/watch?v=XMM66hdD168&feature=plcp

There are some other manufacturers

Although there are often no rebates on portable lighting fixtures, including task lights, the rebates can be indirect with higher rebates on lower wattage ambient lights. Even many people who do not scientifically know which light levels and color tones are best for which parts of the day or task seem to have an innate sense of what should be used when. Tunable undercabinet LED task lights are also being developed. Occupancy sensors can control both types.

THE NEXT STEP

In applications where task lights, especially tunable ones, are well suited, the next step is to also have LED ambient lighting. There are already some very good dimming and Kelvin-changing LED troffers.

The PlanLED Galaxia won a 2012 Next Generation Luminaires Design-recognized award. Now PlanLED and others have better troffers and troffer kits.

Following are some websites for videos on dimming and Kelvin-changing LED troffers. Down the road, if these weblinks no longer work, you should be able to do a YouTube or web search to see some videos.

www.youtube.com/watch?v=OPaKuOeuMg8
www.youtube.com/watch?v=oJKf3kah15E

Figure 18-1 shows a private office space that replaced two 2x4 troffers with one 2x2 PlanLED Galaxia Kelvin-changing troffer and a PlanLED Prism TL4000 desk lamp at Mission Produce Sumner Distribution Center in Washington. While over 70% energy reduction was achieved, user comfort also improved significantly.

My favorite tunable LED desk-mount task light is the PlanLED TL7000. (See Figure 18-2.) It is tall enough be over computer monitors so as to shine light both in front and in back of them to reduce contrast ratios, which can reduce eye strain and headaches. It is also tall enough to mimic an undercabinet task light. The extra

**Figure 18-2. PlanLED
TL7000 desk-mount
task light**

**Figure 18-1. Office with a successful
retrofit resulting in over 70% energy
reduction**

leg allows for long reach, which can be very useful in reading blue prints, maps, etc. The long LED array reduces shadowing compared to many other task lights; it is also low glare. The controls for light level and CCT are easy to understand and use. There is also a timer that automatically turns off the task light after so many minutes. It is representative of the features and performance that, in the near future, will be the standard of the industry.

Table 18-2 is a feasibility table similar to one shown before, but this one has some tunable LED task lights and some tunable LED ambient lights. With those, worker productivity dollar amounts are increased.

OTHER APPLICATIONS

Task ambient lighting can be used for many other applications. One is industrial, where people are working at machines,

Table 18-2.

TYPICAL OFFICE 12' long x 10' wide x 9' high

existing: blended KWH rate $0.15 | additional air conditioning savings (1.00 is none) 1.05 | /KWH saved first year rebate $0.05

application and fixture type: 12 x 10 x 9 office area with 2 2x4 18 cell parabolic troffers, each with 3 32W 735 20,000 hour rated F32T8s and generic .88 BF ballasting (1.51 watts per square foot) — annual hours 3000 — annual electric cost $91.00 — watts 180

option	retrofit option description	rated lamp life hours @ 3 hour cycles	watts (some averaged)	watts per square foot	watts reduction	% watts reduction	annual unit electric cost savings	appr led cost	rebate	per year maintenance savings benefit for comprehensive long term benefit and payback	per year improved worker productivity from improved lighting quality benefit for comprehensive long term benefit and payback	per year combined maintenance savings and improved worker productivity benefit for long term benefit and payback	payback in years just electricity	payback in years electricity & maintenance	payback in years comprehensive	long term benefit just electricity	Long term benefit electricity & maintenance	long term benefit comprehensive
A	Retrofit each troffer with 3 25W F32T8 5000K lamps & .71 BF extra efficient program start parallel wired ballast	30,000 - 36,000	112	0.93	68	38%	$32	$120	$10	$2	$2	$4	3.4	3.4	3.0	$212	$221	$241
B	Retrofit each troffer with upscale kit which eliminates parabolic louvers. 1 high lumen F32T8 5000K lamp & 1.15 BF extra efficient program start ballast	30,000 - 36,000	78	0.65	102	57%	$48	$230	$15	$6	$250	$256	4.5	4.3	0.7	$267	$312	$2,812
B1	Retrofit each troffer with upscale kit which eliminates parabolic louvers. 1 high lumen F32T8 5000K lamp & .89 BF extra efficient program start ballast. Also include 8W dimming & Kelvin changing LED task light.	30,000 - 36,000 for T8s	67	0.56	113	63%	$53	$310	$17	$6	$400	$406	5.5	5.4	0.6	$241	$284	$4,284
B2	Retrofit each troffer with upscale kit which eliminates parabolic louvers. 1 high lumen F32T8 5000K lamp & .71 BF extra efficient program start ballast. Also include 2 8W dimming & Kelvin changing LED task lights.	30,000 - 36,000 for T8s	62	0.52	118	66%	$56	$380	$18	$6	$450	$456	6.5	6.4	0.7	$195	$238	$4,738
C	Remove both troffers. Install 8' suspended indirect/direct fixture that has 2 high lumen F32T8 5000K lamps & 1.15 BF extra efficient program start ballast.	30,000 - 36,000	70	0.58	110	61%	$52	$410	$17	$6	$350	$356	7.6	7.5	1.0	$126	$170	$3,670
C1	Remove both troffers. Install 8' suspended direct/indirect fixture that has 2 high lumen F32T8 5000K lamps & .89 BF extra efficient program start ballast. Also include 8W dimming & Kelvin changing LED task light.	30,000 - 36,000 for T8s	63	0.53	117	65%	$55	$490	$18	$6	$500	$506	8.5	8.4	0.8	$80	$123	$5,123
C2	Remove both troffers. Install 8' suspended direct/indirect fixture that has 2 high lumen F32T8 5000K lamps & .71 BF extra efficient program start ballast. Also include 2 8W dimming & Kelvin changing LED task lights.	30,000 - 36,000 for T8s	60	0.50	120	67%	$57	$560	$18	$6	$550	$556	9.6	9.5	0.9	$25	$67	$5,567
G	Retrofit each troffer with high performance hardwired LED troffer kit with batwing distribution, set at 35W, so sufficient light at end of life.	50,000 - 100,000	70	0.58	110	61%	$52	$270	$17	$12	$250	$262	4.9	4.6	0.8	$266	$370	$2,870
G1	Retrofit each troffer with high performance hardwired LED troffer kit with batwing distribution, set at 30W. Also include 8W dimming & Kelvin changing LED task light.	50,000 - 100,000	60	0.50	120	67%	$57	$340	$18	$12	$400	$412	5.7	5.5	0.7	$245	$347	$4,347
G2	Retrofit each troffer with high performance hardwired LED troffer kit with batwing distribution, set at 25W. Also include 2 8W dimming & Kelvin changing LED task lights.	50,000 - 100,000	55	0.46	125	69%	$59	$420	$19	$12	$450	$462	6.8	6.6	0.8	$189	$291	$4,791
GH	Retrofit each troffer with high performance dimming & Kelvin changing hardwired LED troffer kit with batwing distribution, set at 35W, so sufficient light at end of life.	50,000 - 100,000	70	0.58	110	61%	$52	$370	$17	$12	$350	$362	6.8	6.6	0.9	$166	$270	$3,770
GH1	Retrofit each troffer with high performance dimming & Kelvin changing hardwired LED troffer kit with batwing distribution, set at 30W. Also include 8W dimming & Kelvin changing LED task light.	50,000 - 100,000	60	0.50	120	67%	$57	$440	$18	$12	$500	$512	7.4	7.2	0.7	$145	$247	$5,247
GH2	Retrofit each troffer with high performance dimming & Kelvin changing hardwired LED troffer kit with batwing distribution, set at 25W. Also include 2 8W dimming & Kelvin changing LED task lights.	50,000 - 100,000	55	0.46	125	69%	$59	$520	$19	$12	$550	$562	8.5	8.3	0.8	$89	$191	$5,691

copyright of Stan Walerczyk of Lighting Wizards. www.lightingwizards.com. 1/1/14 version tables, etc. Instead of trying to get all of the light from hibays 20-60 feet high, installing fluorescent, LED hooded industrial, or similar fixtures a few feet over workers' heads could improve lighting and save a ton of energy.

WRAP UP

Since well-designed task ambient lighting is so good, please start using it and allow the DOE to take it off its top ten under-used technologies for saving energy.

Chapter 19

Hibays

INTRODUCTION

Until early 2013, I considered LED high bays (which I call "hibays") really only cost effective for cold storage applications and early adopters. But since the spring of 2013, there have been several LED hibays that can be cost effective for many applications and end users. Although they cost significantly more than high performance fluorescent and electronically ballasted ceramic metal halide (CMH) hibays, these LED hibays can provide higher long-term benefits. So, if the client has and is willing to spend the extra money up front, LED hibays can be a very good solution.

Here are six very good requirements for LED hibays:

- At least 100 LPW out of the fixture in early 2014, at least 120 LPW in 2015, etc.

- At least 80 CRI

- At least 140°F ambient temperature rating in hot applications

- 10-year warranty in most applications

- Qualify for rebates for most applications

- And of course, decent pricing

BEFORE REALLY STARTING

Induction is hardly ever mentioned in this chapter because:

- Only 70-80 LPW with lamp and generator

- Typically bad fixture efficiency because lamp is so large

- Typically bad optical control because lamp is so large

- Although lamps are rated for 100,000 hours, generator may last significantly less

- Mature technology that is becoming more outdated each year compared to LED, etc.

- Customers may not be able to get warranty support or replacement parts down the road

So, if you are considering induction for hibays or any other applications, please reconsider.

LED T8s should not be used in hibays or any other application. LED kits are not recommended either, because it is much better to use LEDs in a new fixture that is designed for the thermal load and light output.

Since high performance fluorescent T8s and electronic ballasts are considerably better than T5HO systems, T8 hibays are the main reference for fluorescent hibays.

SIX REQUIREMENTS FOR LED HIBAYS

You can compare the following requirements with incumbent technologies.

At Least 100 LPW

Like other LED products, this is the LPW out of the hibay. There are several LED hibays with 100 LPW, and by early 2015 that could be up to 120+ LPW.

While high performance fluorescent T8 lamps and ballasts can provide about 100 LPW, if you put them in a hibay, there may only be about 70-90 LPW out of the fixture, because of fixture efficiency and thermal losses. Fluorescent is also a fairly mature technology, and very little effort is being made to improve it.

Ceramic metal halide (CMH) lamps and electronic ballasts can provide 110 LPW, but inside a hibay, the LPW can range from 80-100. This technology is improving, so efficiency will increase. At least one manufacturer is working on the concept of lower wattage replacement lamps, with the same lumens as existing ones, that can work on existing ballasts. For example, if you buy a 210W lamp and ballast now, when the lamp burns out it could possibly be replaced with a 190W lamp. So, wattage could be saved now and down the road.

For all technologies, it is not just about LPW out of the fixture but where those lumens are going. This can be called effective lumens. Since LED hibays can have very good optical control, they can often provide the required light levels with significantly fewer lumens.

Lumen depreciation should also be discussed. High performance fluorescent T8 lamps usually lose only 8-10% of initial lumens over rated life. For CMH it can be about 20%. But, LED may lose 30% at L70, which is the end of rated life. So, an LED system should be somewhat overlit initially or have some kind of constant lumen control system.

At Least 80 CRI

Many LED hibays have only about 70 CRI because they use LEDs that are used for exterior applications; this is about the same CRI as probe or pulse start quartz MH. But to compete with fluorescent and to be closer to CMH, at least 80 CRI is recommended in LED hibays. High performance fluorescent lamps have 80-85 CRI. Most CMH lamps have 90+ CRI, which can be a significant advantage in retail stores, convention centers, car dealerships, etc.

At Least 140°F Ambient Temperature Rating

This is not that important in a conditioned space or one that does not get very hot, but it can be very important in many warehouses and industrial facilities. Often the temperature can be 20-40°F degrees hotter at fixture height than close to the ground. So, if it is 100°F six feet above floor level, it may be 130°F at fixture height. Some LED hibays are only rated for 120°F or 125°F, which is not good in hot applications. Even LED hibays rated for at least 140°F will provide less light at high temperatures, because LEDs get less efficient with heat.

Heat can also dramatically reduce light output of fluorescent lamps, but well-designed fixtures and lamps with amalgam can help. Heat can also significantly reduce electronic ballast life.

With CMH and all HID lamps, temperature is not a concern; they provide about the same amount of light when very hot or when very cold. However, heat can significantly reduce electronic ballast and LED driver life. There are some applications, such as metal foundries, that are too hot for anything electronic. For these applications, CMH or quartz pulse start MH with a magnetic ballast may be best.

10-Year Warranty in Most Applications

For LED hibays, warranty is so much more important than rated life, because warranty is how long the manufacturer will stand behind its product.

Rated life is usually listed at L70 (when the LEDs are expected to maintain 70% of initial lumens). However, various manufacturers use different ways to establish L70. Some manufacturers list a lower hour rating based on no more than six times the tested hours of the LEDs, while others list a longer life based on a calculated or projected approach.

What is very important here is that L70 is only for the LEDs, which may last longer than the driver, electrical connections, exterior finish, etc. So, I do not care that much about rated life numbers but want to know if the manufacturer will warranty the hibay, including LEDs and drivers, for ten years in most applications. Maybe in 24/7 and/or hostile environments, the warranty can be less.

Some LED hibay manufacturers have a standard five-year warranty but will double it to ten years if the customer pays an extra 5-10% initially. A ten-year warranty can really help customers to approve projects. Many end users do not want to approve a project if the payback length happens after the warranty expires. Since the payback can be over five years, the ten-year warranty can really help.

Even if various warranties have the same length, please read them to see what they cover, because there can be significant differences. Often there is no labor credit. For fluorescent and CMH, the warranty is usually 3 years, parts only, for lamps and 5 years parts and labor for electronic ballasts. Some of the extra long life fluorescent lamps with up to a 67,000-hour rating have up to a 5-year warranty.

For any lighting product, especially LED, what is the worth of the warranty? If the company has only been in business for one or two years, a ten-year warranty may only be worth the paper it is written on. With the way LED technology is progressing, even some top name brand companies may not survive for ten years. Often, a third-party insurance policy from a reputable insurance company is the best way to go, because no matter what happens to the manufacturer, customers will still be able to get funds to replace fixtures that failed during the warranty period. Some LED product manufacturers have third-party insurance policies.

Qualify for Rebates

Most of the time it is good to get hibays that qualify for rebates. For LED hibays in most parts of North America, they should be qualified by the DesignLights Consortium (DLC). However, the DLC may have restrictions that do not work out for specific applications (www.designlights.org).

While some rebate providers use the Lighting Design Lab (LDL) for approval of LED hibays, in certain ways the LDL may be better than the DLC (http://lightingdesignlab.com).

Many rebate providers require that fluorescent T8 lamps and electronic ballasts be approved by the Consortium of Energy Efficiency (www.cee1.org).

There do not seem to be very many restrictions with CMH lamps and electronic ballasts.

Pricing

LED hibay pricing has been coming down but is still considerably more than other technologies. For a decent quantity, LED hibays, which can provide as much light as 400W HPS or probe start MH, may cost $350-$600 (in 2014), and that price will be lower in 2015 and beyond because fewer LEDs will be needed and the required LEDs will cost less. There are some economy versions at a lower cost, and some of them may be a good value in some projects.

Good, six-lamp F32T8 hibays may currently cost about $125-$150. There are some imported four-lamp T5HO hibays for as low as $80, but again, sometimes you get what you pay for.

A CMH lamp and electronic ballast system may currently cost about $150, and they can go into existing or new fixtures.

For any hibay, initial pricing is really a small portion of the life cycle cost, or long-term benefit, of ownership.

OTHER CONSIDERATIONS

Rebate Amounts

Numerous rebate programs provide higher rebates for LED than for other technologies. The purpose of these emerging technology rebates is to help increase the volume and reduce the price of LED products. For example, the California investor-owned utilities, on May 1, 2013, began providing a rebate of $.08/kWh saved over the first year for LED products and only $.03/kWh for most everything else.

Controls

LEDs are great with on/off occupancy sensors, because life

is not shortened as with fluorescent lamps. LEDs are much better with dimming than fluorescent, because LEDs can maintain or improve efficiency, while fluorescent systems get considerably less efficient when dimmed. If occupancy sensors will be used with fluorescent hibays, then parallel wired, program start, electronic ballasts should be used for good lamp life.

Electronically ballasted CMH can be dimmed more efficiently than fluorescent because no lamp cathode heating is needed, but CMH lamps can turn greenish if dimmed below 50% (probably only good on St. Patty's Day). CMH can be used with occupancy sensors, but only at bi-level and adjusting down to 50%, not off.

Across the board, numerous companies are pushing fancy control systems, usually wireless. But if you do your homework, you will find that they are often not cost effective. This is because the connected wattage is so low that not very much extra energy can be saved.

Uplight

Uplight is not very important in many applications but is in others. Most LED hibays do not provide any uplight, but more are being developed to do so. Fluorescent hibays can provide uplight with clear prismatic refractors (instead of metal reflectors), vents in reflectors, or lamps on top of the fixtures. CMH is probably the best for uniform and aesthetically pleasing uplight in hibays with clear prismatic acrylic or glass domes.

Financing

Since LED hibays are on the expensive side, some manufacturers offer positive cash flow financing at no or limited up front out of pocket money. Such a plan features monthly savings that are more than the monthly payments to the utility and lending institution. Banks, ESCOs, and some utilities can also provide financing.

Spectrally Enhancing Lighting

Almost all of my fluorescent hibay projects have been with 5000K, and I plan to continue that with LED. This is especially the case since the Illuminating Engineering Society (IES) approved TM-24-13, which details the benefits of high-Kelvin lighting, especially the extra wattage savings. See: www.iesna.org

T8 Hibays

If you are going to get T8 hibays, make sure you get good ones that are specifically designed for T8 lamps, not for the smaller-diameter T5HO lamps. There should also be only one lamp in each reflector cavity, the reflector should be at least nominal 4″ wide, and the bottom of the reflector should be below the bottom of the lamp.

Dirty Environments

Most LED hibays have flat tops where the cooling fins are. If there is a lot of dirt that can settle on top of such fixtures, it can be difficult to keep the LEDs and drivers cool. However, some LED hibays have primarily vertical cooling fins, which are more effective at removing heat because dirt is less prone to collect.

Wet Rated, Hazardous Location, Etc.

For any of these requirements, it is often possible to use LED, fluorescent, and electronically ballasted CMH. Since these fixtures may be expensive, often the best solution in a retrofit project is to install CMH lamps and electronic ballasts in existing HID fixtures.

Where the LED Hibays are Made

Some LED hibays are mainly made in the USA. Others are primarily manufactured in Japan, South Korea, Taiwan, etc., with some being made in China.

Human Centric Lighting

This is lighting that can improve circadian rhythms for better alertness, sleep, mood, etc. Hopefully, there will be some good tunable (dimming and Kelvin-changing) LED hibays soon. Considerable information on human centric lighting is available in Chapter 29.

TABLE TIME

Tables 19-1 through 19-5 compare high performance LED, fluorescent T8, and electronically ballasted CMH hibays. The LED hibays are considered to have at least 100 LPW, 80 CRI, and up to a 10-year warranty. But, please be aware that (in 2013) many LED hibays only have 80-90 LPW, 70 CRI, and only a 5-year warranty.

I prefer end of life (EOL) compared to initial or mean lumens because it is important to know if there will be sufficient light near the end of lamp life (L70 for LED). But, LED hibays have the best initial lumens per watt out of the fixture.

Useful lumens are not only ones directed to useful areas but also those not over-lighting directly underneath fixtures. LED hibays with well-designed optics can save energy by providing uniform lighting without the wasteful blob of light underneath. So, minimum foot-candles or lux between hibays is much better than average foot-candles or lux.

In Tables 19-1 and 19-2, the most important columns are the two far right ones that are bolded. They include the benefits of spectrally enhanced lighting.

Annual hours, kWh, material cost, and labor cost can vary, depending on the utility, rate schedule, quantity, whether union or prevailing wages apply, fixture height, etc. However, the relative comparisons among the options are still valid. Although many people just use payback (with which the LED hibays are not nearly as good), LED is best based on long-term benefit, including maintenance savings. As you can see, the installed cost in early 2014 for the LED highbays is about twice that of T8 hibays.

Figure 19-3 shows an example with an existing T5HO hibay.

Table 19-1

HIBAY LIGHT OUTPUT AND WATTAGE COMPARISON - 2013

hibay & lamp fixture type	CRI	lamp life @ 10 or 12 hour cycles	initial lamp lumens	BF	actual initial lamp lumens	EOL lamp lumen maintenance	EOL lamp lumens	luminaire efficiency	useful lumens factor	EOL useful luminaire lumens	system watts	wattage savings compared to standard MH	EOL useful luminaire lumens per watt	S/P ratio	EOL useful luminaire task modified lumens	EOL useful luminaire task modified lumens per watt
400W 4000K standard MH lamp, magnetic ballast & spun aluminum reflector	70	20,000	36,000	1.00	36,000	45%	16,200	75%	75%	9,113	458		20	1.65	22,403	49
160W 5000K LED hibay (16,000 useful initial lumens out of fixture)	80	100,000+								11,200	160	298	70	2.00	38,976	244
130W 5000K LED hibay (13,000 useful initial lumens out of fixture)	80	100,000+								9,100	130	328	70	2.00	31,668	244
6 3000 lumen F32T8 850 lamps, high performance 1.15 BF program start ballasting & enhanced aluminum reflectors	80	36,000 - 42,000	18,000	1.15	20,700	92%	19,044	90%	75%	12,855	211	247	61	1.95	42,864	203
6 extra long life 2900 lumen F32T8 850 lamps, high performance .89 BF program start ballasting & enhanced aluminum reflectors	80	60,000 - 67,000	17,400	0.89	15,486	92%	14,247	90%	75%	9,617	168	290	57	1.95	32,067	191
210W 4200K CMH lamp & electronic ballast in existing hibay	90+	30,000 - 50,000	24,000	1.00	24,000	80%	19,200	75%	75%	10,800	221	237	49	1.80	31,104	141
210W 4200K CMH lamp, electronic ballast, high performance dome	90+	30,000 - 50,000	24,000	1.00	24,000	80%	19,200	90%	75%	12,960	221	237	59	1.80	37,325	169

Notes
BF stands for ballast factor and EOL stands for end of life.
Fluorescent lamp lumens are based on optimal temperatures & can be adjusted with lumen/temp tables provided by manufacturers.
HID have magnetic ballasts.
Fluorescents have extra efficient program start electronic ballasts.
S/P is scotopic/photopic. Light with more blue content have higher S/P ratios and are perceived brighter by the human eye. This is spectrally enhanced lighting.
End of life useful luminaire task modified lumens = end of life useful luminaire lumens x (S/P).⁸ [.8 exponent]
Prepared by Stan Walerczyk of Lighting Wizards   www.lightingwizards.com   1/1/14   version

Table 19-2

HIBAY COST EFFECTIVENESS COMPARISON - 2014

$0.150	KWH rate				15						years of long term benefit			
existing				proposed											
fixture & application type	unit watts	annual hours	annual elec-trical cost	option letter	retrofit/replacement option description	watts	watts redux	annual elec-trical savings	rebate rate	appr. rebate	appr. installed cost	rated lamp life @ 10-12 hour cycles	pay-back just elec-tricity	long term benefit just elec-tricity	long term benefit including maint-enance savings
warehouse, industrial or big box retail hibay with basic grade 20,000 hour rated 400W probe start MH lamp, magnetic ballast & basic grade spun aluminum dome	458	5500	$378	A	new 160W LED hibay	160	298	$246	$0.08	$131	$750	100,000	2.5	$3,069	$4,069
				B	new 130W LED hibay	130	328	$271	$0.08	$144	$700	100,000	2.1	$3,503	$4,503
				C	6 3000 lumen F32T8 850 lamps, high performance 1.15 BF program start ballasting & enhanced aluminum reflectors	211	247	$204	$0.03	$41	$300	36,000 - 42,000	1.3	$2,797	$3,197
				D	6 extra long life 2900 lumen F32T8 850 lamps, high performance .89 BF program start ballasting & enhanced aluminum reflectors	168	290	$239	$0.03	$48	$300	60,000 - 67,000	1.1	$3,337	$3,837
				E	210W 4200K CMH lamp & electronic ballast in existing hibay	221	237	$196	$0.03	$39	$250	30,000 - 50,000	1.1	$2,722	$3,022
				F	210W 4200K CMH lamp, electronic ballast, high performance dome	221	237	$196	$0.03	$39	$350	30,000 - 50,000	1.6	$2,622	$2,922

copyright of Stan Walerczyk of Lighting Wizards, www.lightingwizards.com, 1/1/14 version

Table 19-3

T5HO HIBAY - 2014

$0.150	KWH rate			option letter	retrofit/replacement option description	proposed (15)						years of long term benefit			
	existing					watts	watts redux	annual elec-trical savings	rebate rate	appr. rebate	appr. installed cost	rated lamp life @ 10-12 hour cycles	pay-back just elec-tricity	long term benefit just elec-tricity	long term benefit including maint-enance savings
fixture & application type	unit watts	annual hours	annual elec-trical cost												
warehouse, industrial or big box retail hibay with 4 54W F54T5HO 841 lamps & generic electronic ballasting	234	5500	$193	A	new 115W LED hibay	115	119	$98	$0.08	$52	$650	100,000	6.1	$875	$1,625
				B	4 47W F54T5HO 850 lamps & high performance program start ballasting	203	31	$26	$0.03	$5	$60	30,000 - 42,000	2.1	$329	$579
				C	new hibay with 6 extra long life 2900 lumen F32T8 850 lamps, high performance .89 BF program start ballasting & enhanced aluminum reflectors	168	66	$54	$0.03	$11	$300	60,000 - 67,000	5.3	$528	$878

copyright of Stan Walerczyk of Lighting Wizards, www.lightingwizards.com, 1/1/14 version

AWARD WINNERS

It is often good to get award-winning LED hibays; however, some manufacturers do not enter their hibays for awards, and there can often be new models which are better than older ones that won awards in 2013. You can get award winners in upcoming years.

Architectural SSL Magazine's
2013 Product Innovation Awards
These include:

• Acuity Brands/Lithonia Lighting I-BEAM LED

• GE/Albeo Technologies HX-Series

• Meteor Lighting/ILOS Corp. 250W High-Ceiling LED Platform

• Lusio Essentials Bay Series

• Hubbell/Columbia Lighting LLHP - 2' Premium High Bay
  (See: www.architecturalssl.com)

Next Generation Luminaires Design Competition
The 2012 awards were presented in early 2013:

• Best in Class - High-bay Industrial Lighting
  — Albeo Technologies' H-Series

• Recognized
  — Digital Lumens' ILE-3-18
  — Lithonia's Proteon
  (See: http://www.ngldc.org/12/indoor/winners.stm)

For various manufacturers and models, you can check LPW, CRI, rated life, warranty, whether qualifying for rebates, pricing, etc.

MANUFACTURERS

Some of the major LED hibay manufacturers are:

- Acuity brands, including Holophane and Lithonia
- Albeo, purchased by GE
- Cooper
- Cree
- Dialight
- Digital Lumens
- Hubbell
- LSI
- Lumenpulse
- Lusio Luminaire
- Philips

There are also numerous smaller companies, including Cool Lumens and PlanLED. PlanLED is noteworthy, because even in 2013 it had a pretty full range of LED hibays. The following can provide a sense of LED performance, pricing, and warranty. Over time, PlanLED and other manufacturers will offer additional product lines which should be better than what is available in mid 2013.

During first quarter of 2014 PlanLED could provide LED hibay solutions with:

- 100+ LPW

- 80+ CRI

- 100,000+ hour rating

- 140°F ambient temperature rating

- Wet location rating

- Motion sensors, photocontrols, and advanced control systems

- Adjustable flood heads

- Uplight

- Powder coating for caustic environments

- DLC and LDL approval, important for rebates

- Up to a 10-year warranty, backed up by a third-party insurance policy

- Very good pricing and conditional direct sales

- Positive cash flow financing

The Luna 130W hibay can usually replace 400W MH or HPS. (See Figure 19-1.) The Luna 99W and 160W hibays are similar.

The AL-480W can replace 1000W MH or HPS. (See Figure 19-2.) This can also be wall mounted, which I recommended for an interior swimming pool for indirect lighting to reduce glare and maintenance costs.

The TM is an adjustable module based high or low bay, which also be used as a flood, wallpack, etc. (See Figure 19-3.)

The KMW120W is an economy LED hibay, which may be cost effective even with relatively low electric rates, rebates, and annual hours of operation. (See Figure 19-4.)

Figure 19-1. Luna 130W hibay

Figure 19-2. AL-480W

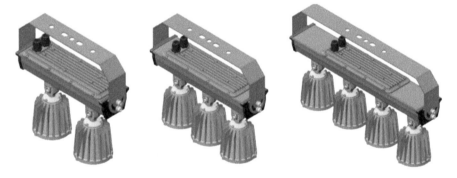

Figure 19-3. TM adjustable module

For PlanLED's Mission case study, see: http://planled. com/?portfolio=3613

CONCLUSION

If the end customer has sufficient funding and is willing to spend it, LED hibays are often the best long-term solution, even by early 2014, and LED hibays will improve in the future. If

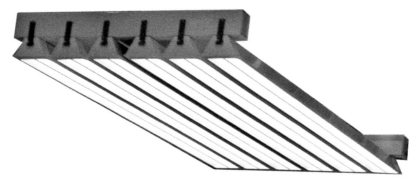

Figure 19-4. The KMW120W

people want the best payback or return on investment (ROI), new T8 hibays or retrofitting with electronically ballasted CMH may be the best in certain projects, at least in early 2014. But, payback is not a very good financial tool because it does not include any benefit after the payback period.

For hibays and other fixture types, more and more end customers with whom I talk want to go with LED, even if there is more initial cost, because they do not want other lighting technologies installed in their buildings that may become outdated in a few years.

Some high tech companies want high tech lighting. Some people want to wait to do anything until LED hibays get lower priced and more efficient, but often the lost savings are never recouped by waiting one, two, or three years.

Many people already know about Acuity Brands, Albeo, Cooper, Cree, Dialight, Hubbell, Philips and some other LED hibay manufacturers, which are all very good. But, please do compare them with some of the smaller companies before you specify or buy.

Chapter 20

Garage Lighting

INTRODUCTION

The garages referred to here are commercial ones, usually multi-floor. They can be totally underground, stand alone above ground, or be a combination of both. There can be a mix of interior and exterior lighting. Often, retrofitting existing fluorescent fixtures with high performance lamps and ballasts, especially with delamping, can be very good. Retrofitting with LED lightbars can also be good. Many times, retrofitting existing well-designed and good condition HID fixtures with electronically ballasted CMH can be very good. This chapter focuses on new LED fixtures.

NEW LED FIXTURES

LED fixtures are becoming quite popular, especially with bi-level motion sensors. By 1Q14 at least 80 LPW is recommended, and some may have 100 LPW or higher.

There are many LED garage fixtures, but several are glare bombs, with the lights looking like MR16 spots when driving and walking in certain directions at certain heights.

Figures 20-1 through 20-3 show low-glare LED garage lights. There may be others. First is the Philips Wide-Lite VizorLED with indirect LEDs (Figure 20-1). Then there is the Kenall TekDek with high performance low glare lens; with the control system there is an L70 signal (Figure 20-2). In the second half of 2013, Cree introduced the VG Series (Figure 20-3).

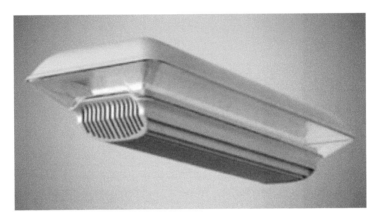

Figure 20-1. Philips Wide-Lite VizorLED garage light

Figure 20-2. Kenall TekDek garage light

Figure 20-3. Cree VG

Cooper's new Top Tier has a WaveStream flat lens with dimple technology from Rambus.

Figure 20-4.

MOTION SENSORS

Bi-level motion sensors can often save considerable kWh cost effectively, especially in company garages where many people come and leave at the about the same time, with very little traffic in between.

However, most of the motion sensors attached to LED garage fixtures are passive infrared, which sometimes work well and sometimes don't. If walking or driving directly toward the sensor, it can take being almost underneath the fixture to go from low to high light levels, which is not good. Mounting sensors in entrances with wired or wireless connection instead of in each fixture may work better.

The next generation of occupancy sensors may use cell

phone camera technology. One example is the National Renewable Energy Laboratory Image Processing Occupancy Sensor (IPOS), which uses parts like those existing in cell phones that have digital cameras. So, no new parts have to be designed and manufactured, but some digital components have to be reprogrammed. Another example is Lighting Science Group's Pixel-View.

Chapter 21

Exterior Lighting

INTRODUCTION

If you have not read the chapters on LED (10) and CMH (9), it may be good to do so, because they are the best two technologies for most exterior applications. Since an entire book could be written on exterior lighting, this chapter aims at a few major issues.

LED

General

By 1Q14 several LED fixtures have at least 100-110 LPW, 100,000+ hour rating, and a ten-year warranty for most applications. LPW will improve later.

Many manufacturers have a five-year warranty, but if the customer pays an extra 5-10% extra up front, the manufacturer will usually double the warranty to ten years, usually a very good value for the customer. Some manufacturers offer a standard ten-year warranty. No matter the warranty length, it is important to read what the warranty covers and does not cover. You may be surprised.

Pricing has and should continue to come down. Even in 2012, some good LED cobraheads with ten-year warranties, replacing 100W HPS, were sold for less than $200 on some good-sized projects.

LED fixtures with well-designed optics can provide sufficient light between fixtures and to desired perimeters without the excessive and wasteful blob of light underneath fixtures. So, minimum foot-candles or lux is much better than average foot-candles or lux. If there are sufficient minimum foot-candles or lux between fixtures and at desired perimeters, there will almost always be sufficient average foot-candles or lux.

Many LED fixtures are dark sky friendly and can prevent light trespass. Many people do not want to go over 4000K, which is about the same CCT as MH and moonlight; however, higher-Kelvin LEDs usually do have more lumens and LPW. Some people prefer 3500K or lower, even if lumens and LPW are worse. It is a matter of choice. There are also complex issues with solar observatories and certain animal species.

Numerous companies offer LED kits for the existing incumbent technology of exterior fixtures. Although some of those may work well in some fixtures in some applications, they are generally not recommended because those fixtures were not designed for the thermal requirements of LEDs and drivers—and, if the existing lens is kept, optical control will not be that good. Some come with their own lenses, which can allow for good optical control.

However, there are some exceptions. Numerous exterior fixture manufacturers are now including LED options in their existing exterior fixture models and offer LED kits for their older, similar fixtures. Since those LED kits are designed for these fixtures, either new or old, with regard to heat and optics these kits may be a very good solution.

LED was often not that good for replacing 400 and higher wattage HPS or MH because so many LEDs are needed to provide sufficient light, which jacks up the cost. By early 2014, there are already LED fixtures that even replace up to 1500W HPS and LED cost effectively.

Good Applications

For LED to be cost effective, at least a good number of the following conditions are necessary.

•   At least $.10/kWh

- At least 4000 annual hours of operation (but continuous is much better)

- Significant rebates

- Existing mercury vapor

- Existing relatively low wattage MH, because neither efficiency nor life is that great

- Existing relatively low wattage HPS because efficiency is not that great

- People do not like the yellow color and low CRI of HPS

- Existing ballasts are old and would need to be replaced soon

- Existing fixtures are not in very good shape

- Need to be dark sky friendly

- Need to stop light trespass

- New construction when need to buy a new fixture anyway

If you need to buy a fixture, LED is often a very good choice, because the cost and benefits of LED can be compared to a base grade fixture. Sometimes the entire cost of the LED fixture is okay to use.

You may find my parking lot project of interest. A company tore down a building and planned to install a parking lot. High performance shoebox pole fixtures with pulse start MH and magnetic ballasts were compared with LED pole fixtures. With the good optics, 20% fewer LED fixtures were needed. That also meant 20% fewer poles, concrete foundations, etc. were needed. The LED option was less expensive based just on initial parts and labor costs, not even including electrical and maintenance savings.

In August of 2013, Cree listed at $99 its XSPR LED cobraheads on its website, which was an attention getter, even though that price is just for a 25W version that can replace 70W HPS for quantities over 500, without photocell or utility setup. Pricing may be $145-$200 for ones that can replace 100W HPS. (See Figure 21-1.) In early 2014, Leotek introduced their equivalent. Other manufacturers will probably do something similar.

Figure 21-1. Cree XSPR LED cobrahead

CMH

As mentioned in the CMH chapter (9), this can be a very good solution. For example, the City of Chicago compared LED, induction, and CMH and decided to go with electronically ballasted CMH in its streetlights. Since there are high wattage CHM systems, these can often be better than LED replacing 400 and above wattage HPS and MH.

LEP

Light emitting plasma may (LEP) be a good solution to replace high wattage HPS and MH, but it is not known whether LEP will remain viable in the long term. Two major manufacturers are Luxim and Topanga Technologies.

INDUCTION

As the induction chapter (8) states, it is best not to even consider induction.

APPLES TO APPLES COMPARISONS

Many sales people and others often compare a brand new LED, CMH, LEP, or induction fixture or kit with existing fixtures, which typically have old lamps, dirty reflectors, and dirty lenses; the samples often look good in comparison. But, as the LED, CMH, LEP or induction systems age with lumen and dirt depreciation, they can often provide insufficient light. These light sources can lose 20-35% of initial lumens at end of life. Plus, there is dirt depreciation. So, it is recommended to compare the new samples with existing fixtures that have new lamps and clean reflectors and lenses. In such an apples to apples comparison, the new samples should provide as least as much light as the existing lighting, unless the existing is overlit.

Chapter 22

Controls

INTRODUCTION

This chapter will not cover control in depth, as entire books have been written on controls. I present four-hour classes through AEE and other organizations that keep evolving, especially with respect to wireless controls.

Although controls can often be very cost effective, this chapter focuses on the fact that automatic controls are often way over-promoted and mandated for retrofits. Controls may be evolving as fast as LEDs. Just like there are so many new LED product manufacturers, there are also many new controls manufacturers. Many of these LED product and controls companies have a limited start.

"UP TO 80%" SAVINGS

You may see the same marketing material that I do, stating savings up to 80% with this or that manufacturer's controls. Although that looks good, it is not realistic. Savings are typically very much less.

Lighting retrofit projects could also save up to 80%, but that is only if there are a lot of T12 lamps, magnetic ballasts, and mercury vapor and high wattage incandescents, which is quite unusual.

STORY

I remember something said years ago when I worked for a lighting retrofit contractor. After I gave a facility manager a proposal showing that a lighting retrofit project would save him 50% on his electric bills from lighting, he said that he had recently received a proposal for occupancy and other controls that promised to cut down 50% of his kWh from lighting. But, of course he knew that if he did both, he would not have a zero electric bill for lighting. This is an example of a case where what is done secondly has diminishing returns because there is less to save.

REDUCE WATTAGE FIRST

It is usually better to bring down wattage (which directly translates to power density) first and then check to see if automatic controls are cost effective. Usually more energy can be saved with relatively low cost lighting products than with automatic controls. Often it is a 2:1 ratio, and that is why control manufacturers are trying to cut their prices in half.

Let's look at a typical example, a 10′ x 12′ private office:

- Original
    — Two 2x4 lensed troffers had three basic grade 32W F32T8 735 lamps and a generic .88 BF instant start electronic ballast
    — 178W
    — 1.48 WSF
    — 2500 annual hours
    — $.15/kWh
    — $66.75 annual consumption

- Lighting retrofit
    — Retrofit each 2x4 troffer with a two-cove white reflector, one high lumen 32W fluorescent F32T8 850 lamp, and high performance 1.00 BF instant start electronic ballast, which is 33W per troffer or lower wattage hardwired LED troffer kit
    — 66W total

- — .55 WSF
- — 112W saved total
- — $120 installed cost
- — $ 20 rebate
- — $100 installed cost after rebate
- — $24.75 annual consumption
- — $.21 annual consumption per square foot
- — $42.00 annual savings
- — 2.4 year payback, based on just hard savings
- — 63% kWh reduction

- Occupancy sensor retrofit
  - — 20% kWh savings from sensor, which is high for many offices
  - — $4.95 annual savings from sensor
  - — $75 for parts and labor for sensor
  - — $10 rebate for sensor
  - — 13.1 year payback

WHO WANTS A SENSOR PROJECT
WITH A 13-YEAR PAYBACK?

Let's look an occupancy sensor installation without retrofitting the lighting. The payback would be 4.9 years with keeping old lamps and ballasts, which would have to be replaced when they burn out (which may be soon). Let's say that the lighting and sensor are done together. The payback would be 3.5 years, which may not be good enough for a lot of end customers who only want less than 3-year paybacks.

With limited funds, it is often better doing additional lighting retrofits than including controls, but there are exceptions. Here are two examples: (a) In bookrack aisles in a university library, I walked through some spider webs between book stacks; (b) I strolled some warehouse rack aisles, leaving footprints in the dust without seeing any other prints. Occupancy sensors may save 95% kWh in these very uncommon applications.

MANUAL CONTROLS

Many controls manufacturers, ASHRAE, California Title 24, and other big brother and ivory tower people "experts" have been pushing automatic controls. However, manual controls (mainly switches) are often the most cost effective and sometimes save the most energy.

Also, there may be some lighting professionals whom you think are independent but are getting direct or indirect compensation from control and/or dimming manufacturers. Maybe background checks and lie detector tests could be done!

There are numerous office workers, who are very diligent in turning off lights when they leave during the day and at the end of the workday. If occupancy sensors are installed, most of these people probably would allow the sensors' 12-15 minute delay before lights go off. So, it is entirely possible that more energy could be used with occupancy sensors. For example, a while back I worked for ESCOs on some public school projects. Data loggers were installed before and after. In a significant number of elementary schools where one teacher basically stays in one classroom most of the day and/or there are energy police students responsible for turning off lights, annual hours that the lights were on actually increased after occupancy sensors were installed because the teachers and students allowed the 12-15 minute delay.

I have consulted for one bio-pharmaceutical company whose sustainability group developed competition among buildings. Workers in buildings who were the most energy efficient in a certain month got free lunches or other perks. Even if office workers are not currently diligent, usually education and motivation (which can be as simple as email reminders a few times a year) may work wonders in private spaces. Such strategies can often be much more cost effective than purchasing and installing occupancy sensors.

WHERE OCCUPANCY SENSORS CAN BE COST EFFECTIVE

Rules of Thumb

In typical applications, I have found that a wall-mounted occupancy sensor needs to control at least 100W, and sometimes at least 150W, to be cost effective. But, with high performance lighting the lighting wattage is usually considerably less than 100W. For example, the previously described private office has 66W after the lighting retrofit.

Ceiling sensors often need to control at least 200W, and sometimes at least 300W, to be cost effective. That is because parts and labor for ceiling sensors can be over $100. However, with high performance lighting, the lighting wattage is often less than 200W.

Fixture-mounted occupancy sensors can often be cost effective in fluorescent or LED hibays, etc.

Owned vs. Non-owned Spaces

Owned spaces are spaces where people feel ownership, including private offices and classrooms in which a teacher stays most of the day. Non-owned spaces include open offices, classrooms where teachers come and go, conference rooms, break rooms, restrooms, and copy rooms. Since people do not feel ownership, they often leave lights on when they leave if the lights were on when they entered. Non-owned spaces are usually much better for occupancy sensors than owned spaces.

USING BREAKERS AS SWITCHES

In some industrial buildings, warehouses, gyms, and other facilities, breakers are used to turn lights on and off every day. That is not good for regular breakers because it can cause them not to automatically turn when there is a safety issue. There are two good solutions. One is using switch-rated breakers. Another is installing a switch or switch box with low voltage relays to the circuit breaker box.

MANDATING CONTROLS

ASHRAE, California's Title 24, and other organizations have started mandating occupancy or vacancy sensors and other controls. Although this may not be bad for new construction, it is ridiculous for retrofits because:

- kWh can actually increase with controls in numerous projects.

- Even when controls save energy, the financial returns can usually get worse.

- Sometimes the financial returns can be bad enough that the end customer does nothing, so no energy is saved.

The big question is, do we benefit from Orwellian Big Brothers, who think they know better than lighting professionals and end customers on specific projects? If some things are cost effective with or without rebates, lighting professionals and end customers will probably use them, but if some products are not cost effective, they will probably not be used. Let the market decide.

ADVANCED CONTROLS

There are a few advanced control technologies that I would like to mention. One is the new generation of wireless controls, which have numerous benefits but may cost $60-$80 per fixture, including the module in the fixture, local transceiver, computer system, internet connection, installation labor, and commissioning labor. Even at $40 per fixture, the cost effectiveness is questionable with very efficient lighting.

Cell phone app controls will probably proliferate. If you do not already know about it, check out the Philips Hue, which is a dimming and color-changing, screw-in LED bulb controlled by a cell phone. (See Figure 11 in Chapter 13.)

The next generation of occupancy or vacancy sensors may use cell phone camera technology. One example is the National Renewable Energy Laboratory Image Processing Occupancy Sensor (IPOS), which uses parts like those existing in cell phones that have digital cameras. So, no new parts have to be designed and manufactured, but some digital components have to be reprogrammed. Another example is Lighting Science Group's Pixel-View.

DIMMING FLUORESCENT BALLASTS

Although numerous dimming fluorescent ballast manufacturers, control companies, and sales people push continuous or staged dimming fluorescent ballasts, they are really expensive energy hogs. Here are some reasons:

- Cost 2-5 times more than high performance fixed output ballasts

- Usually less efficient than high performance fixed output ballasts

- Really get inefficient at .70 BF and below because the ballasts need to heat lamp cathodes so that lamps do not flicker, spiral, or turn off

- For example, dimming every dimming ballast to 50% light level consumes 20-30% more wattage than turning off every other equivalent fixed output ballast.

- Most dimming ballasts are series wired, so if one lamp burns out, at least one other lamp also goes out, which increases maintenance costs.

- DALI or similar digital dimming ballasts can consume up to .75W when they are turned off. In a building with 1000 ballasts, that is 750W, which is substantial.

Even if dimming ballasts and controls, such as photocontrols, save energy, this is often not cost effective compared to high performance fixed output fluorescent systems, which can provide very low power densities (such as 66W and .55 WSF in the 120 square foot private office previously described). Even 40% savings would only be $9.90 per year. With two dimming ballasts, a control system, and labor, the cost could be $160 after rebate, which would provide a 16.2 year payback—actually a longer one because the ballasts and the control system would probably have to be replaced in that time.

Please be aware that fixed output ballasts can be used for daylight harvesting. Some or all of them could be turned off manually or automatically when there is sufficient daylight.

The future of dimming will be with LED because LED systems can maintain or improve efficiency when dimmed, and dimming LED drivers do not cost that much more than standard drivers. But, LED systems can usually provide a lower power density and electric cost than fluorescent, so it will be necessary to check as to whether dimming LED systems are cost effective on specific projects.

WILL CONTROLS COMPANIES BE AROUND DOWN THE ROAD?

Controls may be evolving as fast as LEDs. Just like there are so many new LED product manufacturers, there are also many new controls manufacturers. Many of these LED product and controls companies have limited start-up funding, and their goal may be to be bought by a large company because, if they are not bought, they may go out of business. If such companies have proprietary systems and go out of business, the end customer may have to replace the entire system.

Chapter 23

Retrofit

INTRODUCTION

This chapter is for those of you who are considering getting or providing a lighting retrofit project. It is also written for the end customer perspective but applies to others.

Well-designed and good-priced lighting projects can be a win-win-win-win for end customers, lighting professionals, utilities and the earth. But there are also many lighting retrofits that, putting it nicely, have been wasted opportunities.

DELAMP OR LAMP-FOR-LAMP RETROFITS

As the delamp or lamp-for-lamp retrofit chapter (7) states, several lighting retrofit contractors and ESCOs like lamp-for-lamp retrofits with 28W or 25W F32T8s because they do not need to measure, order, have delivered, carry around, and install kits, while hoping they are the correct size. Although this lamp-for-lamp strategy is easy and may provide the best payback, it does not save the most energy and does not provide the best long-term benefits.

FEASIBILITY STUDIES

Although some contractors and ESCOs provide one high option and one low cost option, many provide only one solution, and sometimes the specifications are not even listed.

Time and time again, retrofit contractors and lighting providers across the country tell me that their end users are cheap. I ask them how they know that, and they just tell me they know. Then I ask them if they have provided higher cost options, which can save more, or better term financials, and they say no.

I almost always provide feasibility studies based on the one or two most common fixture types in a project. Then I go over the pros and cons of each option with the client. Although my clients, who are usually end users, may not go with the most expensive option, they almost always go with something other than the least expensive one because they understand the value of higher-priced solutions that provide better quality light, more savings, and better long-term benefits.

Table 23-1 is an example with existing parabolic troffers. Many retrofit contractors and ESCOs push option A, which is lamp-for-lamp retrofit with reduced wattage lamps and new ballasts, because it is easy and the straight payback is good. However, none of my clients have selected that because they understand that there are much better solutions. They have at least gone with B or B1 and now, with the latest generation of hardwired LED troffer kits, I expect that many of them will go with at least G or G1. If they really understand human centric lighting, then GH, GH1, or GH2 would be the best.

PROPOSALS

Some contractors provide lighting retrofit proposals that are incomplete. Such proposals may just have total cost before and after rebates or total savings and payback without any breakdown (number of each type of fixtures, specified products, how savings and rebates were calculated).

Since I used to work for lighting retrofit contractors and still have several contractor friends, I am aware that this is often done because it takes very little time, and because competitors cannot provide an apples to apples bid or show how their recommendations are better.

End customers should demand complete proposals, including:

Table 23-1

TYPICAL OFFICE 12' long x 10' wide x 9' high

option	retrofit option description	rated lamp life hours @ 3 hour cycles	watts (some averaged)	watts per square foot	watts reduction	% watts reduction	annual unit electric cost savings	appr unit install-led cost	rebate	per year maintenance benefit (10)	per year improved worker productivity benefit	per year combined savings & productivity	payback in years just electricity	payback in years electricity & maintenance	payback in years comprehensive	long term benefit just electricity	Long term benefit electricity & maintenance	long term benefit comprehensive
A	Retrofit each troffer with 3 25W F32T8 5000K lamps & .71 BF extra efficient program start parallel wired ballast	30,000-36,000	112	0.93	68	38%	$32	$120	$10	$2	$2	$4	3.4	3.4	3.0	$212	$221	$241
B	Retrofit each troffer with upscale kit which eliminates parabolic louvers, 1 high lumen F32T8 5000K lamp & 1.15 BF extra efficient program start ballast	30,000-36,000	78	0.65	102	57%	$48	$230	$15	$6	$250	$256	4.5	4.3	0.7	$267	$312	$2,812
B1	Retrofit each troffer with upscale kit which eliminates parabolic louvers, 1 high lumen F32T8 5000K lamp & .89 BF extra efficient program start ballast. Also include 8W dimming & Kelvin changing LED task light.	30,000-36,000 for T8s	67	0.56	113	63%	$53	$310	$17	$6	$400	$406	5.5	5.4	0.6	$241	$284	$4,284
B2	Retrofit each troffer with upscale kit which eliminates parabolic louvers, 1 high lumen F32T8 5000K lamp & .71 BF extra efficient program start ballast. Also include 2 8W dimming & Kelvin changing LED task lights.	30,000-36,000 for T8s	62	0.52	118	66%	$56	$380	$18	$6	$450	$456	6.5	6.4	0.7	$195	$238	$4,738
C	Remove both troffers. Install 8' suspended indirect/direct fixture that has 2 high lumen F32T8 5000K lamps & 1.15 BF extra efficient program start ballast.	30,000-36,000	70	0.58	110	61%	$52	$410	$17	$6	$350	$356	7.6	7.5	1.0	$126	$170	$3,670
C1	Remove both troffers. Install 8' suspended direct/indirect fixture that has 2 high lumen F32T8 5000K lamps & .89 BF extra efficient program start ballast. Also include 8W dimming & Kelvin changing LED task light.	30,000-36,000	63	0.53	117	65%	$55	$490	$18	$6	$500	$506	8.5	8.4	0.8	$80	$123	$5,123
C2	Remove both troffers. Install 8' suspended direct/indirect fixture that has 2 high lumen F32T8 5000K lamps & .71 BF extra efficient program start ballast. Also include 2 8W dimming & Kelvin changing LED task lights.	30,000-36,000 for T8s	60	0.50	120	67%	$57	$560	$18	$6	$550	$556	9.6	9.5	0.9	$25	$67	$5,567
G	Retrofit each troffer with high performance hardwired LED troffer kit with batwing distribution, set at 35W, so sufficient light at end of life.	50,000-100,000	70	0.58	110	61%	$52	$270	$17	$12	$250	$262	4.9	4.6	0.8	$266	$370	$2,870
G1	Retrofit each troffer with high performance hardwired LED troffer kit with batwing distribution, set at 30W. Also include 8W dimming & Kelvin changing LED task light.	50,000-100,000	60	0.50	120	67%	$57	$340	$18	$12	$400	$412	5.7	5.5	0.7	$245	$347	$4,347
G2	Retrofit each troffer with high performance hardwired LED troffer kit with batwing distribution, set at 25W. Also include 2 8W dimming & Kelvin changing LED task lights.	50,000-100,000	55	0.46	125	69%	$59	$420	$19	$12	$450	$462	6.8	6.6	0.8	$189	$291	$4,791
GH	Retrofit each troffer with high performance dimming & Kelvin changing hardwired LED troffer kit with batwing distribution, set at 35W, so sufficient light at end of life.	50,000-100,000	70	0.58	110	61%	$52	$370	$17	$12	$350	$362	6.8	6.6	0.9	$166	$270	$3,770
GH1	Retrofit each troffer with high performance dimming & Kelvin changing hardwired LED troffer kit with batwing distribution, set at 30W. Also include 8W dimming & Kelvin changing LED task light.	50,000-100,000	60	0.50	120	67%	$57	$440	$18	$12	$500	$512	7.4	7.2	0.7	$145	$247	$5,247
GH2	Retrofit each troffer with high performance dimming & Kelvin changing hardwired LED troffer kit with batwing distribution, set at 25W. Also include 2 8W dimming & Kelvin changing LED task lights.	50,000-100,000	55	0.46	125	69%	$59	$520	$19	$12	$550	$562	8.5	8.3	0.8	$89	$191	$5,691

$0.15 blended KWH rate | 1.05 | additional air conditioning savings (1.00 is none) | $0.05 /KWH saved first year rebate | $0.05

existing: application and fixture type — 12 x 10 x 9 office area with 2 2x4 18 cell parabolic troffers, each with 3 32W 735 20,000 hour rated F32T8s and generic 88 BF ballasting (1.51 watts per square foot); watts 180; annual hours 3000; annual electric cost $81.00

copyright of Stan Walerczyk of Lighting Wizards, www.lightingwizards.com. 1/1/14 version

- kWh rate, perhaps with a break down of peak and normal load

- Existing number of each type of fixture

- Existing wattage of each of fixture

- Existing annual hours of each fixture in each room

- Existing kWh of each fixture in each room

- Specific recommendation for maintaining existing fixtures, fixture for fixture replacement, removing some fixtures, or a complete new layout

- Proposed unit wattage, kWh, cost, and rebates

- Total KW, kWh, cost, and rebates

- Unit financials, including something better than payback and ROI

- Total financials, including something better than payback and ROI

- Room-by-room listings with corresponding numbered map

For this extra work, it is not unreasonable for end customers to pay preferred contractors to do this on certain projects.

APPLES TO APPLES COMPARISON

Instead of different bids from different contractors, it is usually best to develop specifcations and have contractors bid apples to apples. Some end users can develop the specs and bid documents themselves. Some end users can use specs from one contractor, pay that contractor a fee, and have that and other contractors bid on those specs.

Some end users hire an independent lighting consultant such as John Fetters at Effective Lighting Solutions, or myself, to develop the specs and other bid documents for contractors to bid on.

John and I can also provide additional services, including leading the bid walk, deciding if equivalents are really equivalents, checking references, taking care of check list items, etc. (Although there may be more, I am not aware of any other good, independent energy efficient lighting retrofit consultants other than John and myself.)

THIRD-PARTY REVIEW

Many end customers can evaluate divergent proposals and pick what they want. Other end customers really do not know enough about lighting and will benefit from hiring someone who can review one proposal, or divergent ones, and tell them which may be a good proposal (if there is one). Again, John Fetters, I, and perhaps somebody else could do that.

POSITIVE CASH FLOW FINANCING

This is described in the Financial Tools chapter (24). Positive cash flow financing can be good, but be sure you know the interest rate and terms.

Chapter 24

Financial Tools

INTRODUCTION

Although many people use payback, there are much better financial tools. Just using hard savings is very limiting. There are numerous ways to fund projects.

PAYBACK

The common definition of payback that most people accept (which includes no benefit after payback) is:

Payback = (initial cost − rebate, if there is one)/1st year savings

But be careful about any manufacturer or other lighting vendor who may plays games with the concept of "payback," including benefits after the payback period. For example, If you see something like 1.5-year payback without including rebate replacing fluorescent F32T8s one for one with LED light bars, that number may help increase sales, but it may not be real payback. That 1.5-year "payback" may be over a 10-year real payback. Let's look at an application with higher than average annual hours of operation and electric rate:

Existing
- 2x4 troffer with 3 fluorescent 32W F32T8 lamps and generic .88 BF instant start electronic ballast
- 89W
- 4000 annual hours
- $.15/kWh electric rate

- $53.40 annual electric cost (89/1000 x 4000 x $.15)

LED light bar retrofit option
- Three 4' LED light bars retrofit
- 66W
- 23W saved (89-66)
- $13.80 annual electric savings (23/1000 x 4000 x $.15)
- $160 installed cost
- 11.6 year payback ($160.00/$13.80)

Even delamping from 3 lamps to 2 LED light bars, which would consume 44W and cost $110, the payback would still be 4.1 years. Let's compare that to retrofitting this fixture with 2 fluorescent high lumen 32W 5000K F32T8 lamps, .77 BF high performance instant start electronic ballast, and a 2-cove white reflector, which consumes 48W and costs $60. The payback would be 2.4 years, a whole lot better than 4.1.

Let's go one step further with payback of 2 LED light bars compared to this fluorescent option:

- 4W saved = (48-44)
- $2.40 annual electric savings (4/1000 x 4000 x $.15)
- $50 additional cost ($110-$60)
- 20.8 years (May be more like 15 years, including having to replace fluorescent lamps and ballasts.)

What is very important is that most people considering LED will probably also consider high performance fluorescent in retrofits, remodels, and new construction.

In general, payback and its inverse, ROI, are really not good financial tools because they do not include any benefit after the payback period. Especially with long life LED products, there is usually much more benefit after the payback period.

Payback is actually risk management, and I really like this quote from pages 2-3 in Chapter 25 of the *9th Edition of the IES Handbook*. (Emphasized text is mine.)

Those who like the simple payback method argue that it is easy to use and a simple way to determine the profitability of a proposed action. In fact, however, it is actually a risk assessment tool posing as a profitability metric. This is seen by examining the question the method answers. It does not answer the question, is a certain investment profitable? Rather, it responds to the concerns of the person who is unsure about the future and hopes to recoup the investment as soon as possible. *If getting money back is the primary concern, then there is no reason to make the investment at all. If no investment is made, then that money is available immediately, and the payback is zero years, the ideal result of a simple payback calculation.*

Often, the payback for an LED solution may be two or three years longer than with incumbent technologies, but the long-term benefits with the LED solution are usually much better. Here is a simple example with two options:

- Option A costs $100,000 after rebate and saves $50,000 per year, which is a 2-year payback, providing $650,000 over 15 years, typical ballast life.

- Option B costs $300,000 after rebate and saves $100,000 per year, which is a 3-year payback, providing $1,200,000 over 15 years.

Although option B costs three times more, it saves twice as much year after year, so the long-term benefit this way is nearly twice as much. Even factoring in the cost of money over time, option B is usually the better solution.

One facet of the cost of money is that if the end customer does not invest into an energy efficient lighting project, that money can be invested into the company, bank, or stock market, any of which should provide income. That income can be compared to the savings from a lighting retrofit project.

A related term is hurdle rate. Some companies, especially

manufacturers, will only purchase new equipment if the profit to cost ratio is high enough. Lighting vendors can ask these companies what their hurdle rate is. Usually the lighting retrofit will well surpass the core business hurdle rate.

I often ask clients whether they really want the shortest payback or the most money in their pocket after so many years. That usually gets them to look at the bigger picture. Some type of long-term financial tool, such as life cycle costing, cost of ownership, savings to investment ratio, modified internal rate of return, etc. is much better than payback.

Any manufacturer calling their calculations payback and including long-term benefits should instead call that some kind of long-term benefit, not "payback."

The only time payback may be good is if a company has a lease, for example three years. If the payback is not less than three years, the lighting retrofit project is probably not cost effective unless the owner is willing to assist.

If a lighting vendor states something like a six month ROI, it's probably a good idea to find another vendor, because that vendor does not understand the simple fact that payback is time and ROI is a percentage. A six-month payback is a 50% ROI.

SOFT SAVINGS

Way too many vendors and end users only include hard savings (which are electrical savings), rebates, and maybe avoided parts maintenance costs. But soft savings (which include improved worker productivity, student test scores, and retail sales) usually dwarf hard savings. Although soft savings can be difficult to quantify, that can be accomplished.

Well before tunable LED products, improving the light quality was often cost effective. Here is one example. Over the years, I would bring clients into the Pacific Gas & Electric Pacific Energy Center's lighting lab in San Francisco.

- If they had lensed troffers in their offices, I turned them on; if they had parabolic troffers in their offices, I turned them on. I would let them fake work for 5-10 minutes.

- I then turned on suspended direct-indirect fixtures and let them fake work for another 5-10 minutes.

- Next I asked them which they liked better. The answer was always the indirect-directs. I asked them how much less time they and their co-workers would waste per day with the better light. They raised their hands and stated 10, 15, 20, and even 30 minutes per day.

Being very conservative, I would use 5 minutes per day to show: If each office worker makes $50,000 per year, that is a $500 annual benefit per person, year after year. That annual $500 benefit dwarfs parts, labor, energy savings, rebate, and maintenance savings.

After going through this experience, one of Brian Liebel's and my clients decided to replace 2x4 parabolic troffers with suspended indirect-direct fixtures, even though the payback based on energy savings was 18 years, because they understood the real value of good lighting. The suspended solution also used less wattage.

Now, being able to include the comprehensive human centric lighting benefits from tunable LED systems, the soft savings could easily at least double, but the customer will need to accept certain soft savings amounts. Soft savings often need to be seen and experienced to be believed. Get some samples installed and evaluate how people like them. Human resource offices are often a very good place for this because they understand the value of worker satisfaction and productivity.

FUNDING

Sometimes end users tell me they would like to do an energy efficient lighting project but don't have the money. I tell them they do have the money but are paying the utility too much of it.

Positive cash flow financing is one example of what I mean. With it, every month the electric savings are greater than the payment to the lighting manufacturer, finance company, bank, or ESCO. So, with no or little out of pocket money, an end customer can actually make money each month. Plus, when the project is paid for, the end customer gets all of the savings.

Since LED products are on the expensive side, and manufacturers of those products want sales, several of them, including PlanLED, are willing to provide financing on many projects.

The large California investor-owned utilities have on-bill financing at zero interest, and that will probably continue. A qualified customer signs up and gets the project installed. The monthly electric bills stay the same, with the savings paying off the consulting, parts, and labor costs of the project. When all of that is paid, the customer's bill goes down. Here is the Pacific Gas & Electric website with information on its program: www.pge.com/mybusiness/energysavingsrebates/rebatesincentives/taxcredit/onbillfinancing

Hopefully, other utilities will offer similar financing.

WHAT ABOUT WAITING?

Some end users want to wait one or even a few years for LED products to become more efficient and lower in cost before doing a retrofit. But, if you do the math, often the lost savings by waiting are never recouped.

JUST GET USED TO IT

Buying LED products can be similar to buying a smart phone, tablet, or computer. You can do a lot of research to find the best personal electronic device, but as soon as you get one, there is often a better one out within a month or so. The same is often the case with specifying and buying LED projects, so just get used to it.

Chapter 25

Rebates

INTRODUCTION

Rebates can be important, but solutions should not be selected on highest rebates. Often the best solutions get lower rebates or do not qualify for them.

GENERAL

For energy efficient lighting and controls, there are basically two types of rebates, also called incentives. There is one, often called prescriptive, in which the rebate amount is fixed, no matter the exact wattage and annual hours of operation. The other type can be called customized, because it is based on kWh saved over the first year. With the latter type there may also be an additional kicker for peak load saved.

Some providers offer what can be called a higher prescriptive or customized rebate for emerging technologies, such as certain LED products. The purpose is to help increase volume and reduce pricing.

There are rebates for some end customers in certain areas but not for others in other areas. It depends on the local utility, which may or not provide rebates, or whether there is a separate organization that provides rebates. Some large investor-owned utilities that provide direct rebates also have third-party rebate programs. The utilities contract these programs out to specific companies to better serve a certain niche of clients and/or product types.

It can sometimes be difficult to find information on rebates and which ones are best. One good source is the Database of State Incentives for Renewables & Efficiency (DSIRE), funded by the U.S. Department of Energy (www.dsireusa.org). RealWinWin is a private company that can assist finding rebate programs, processing rebate applications, etc. (www.realwinwin.com).

If there are rebates, and if the end customer wants them, it is very important to make sure to get the energy efficient products qualifying for rebates. Too often somebody does a retrofit with basic grade T8s and generic electronic ballasts that a rebate organization allows, and nobody is happy.

Many rebate organizations require T8 lamps and electronic ballasts to be approved by CEE. The 32W F32T8s have to be high lumen, high CRI, and long life. High performance 28W and 25W T8s are also approved. Fixed output and dimming electronic ballasts for T8s are required (www.cee1.org).

For residential grade LED products, including MR16 and PAR lamps, which are also used in retail stores and other commercial applications, many rebate organizations require Energy Star approval (www.energystar.gov).

For commercial grade LED products, many rebate organizations require DLC approval (http://www.designlights. org).

Some rebate programs in the Pacific Northwest and maybe other locations allow approval from the Lighting Design Lab (http://lightingdesignlab.com).

Often retrofitters and end customers want to go with the solution having the highest rebate; however, often the option with the lower rebate is actually better for numerous projects. An example is that some providers provide a higher rebate for 28W and 25W fluorescent F32T8s compared to high lumen 32W fluorescent F32T8s, even if the 32 watters can save more energy with delamping or lower BF ballasts. Fewer lamps mean fewer lamps to buy, stock, replace, and recycle now and down the road.

Sometimes higher rebates for some products is good. But, in general, I wish rebates were the same for all products, based on how much energy is saved.

Sometimes it is good to use non-rebatable products on purpose. Many rebate programs do not allow extra long life T8s because they do not have sufficient lumens. But, with up to 67,000 rated hours at 12-hour cycles with program start ballasts, those lamps can really reduce maintenance costs in many applications.

For at least through mid 2013, the DLC has been inhibiting LED in retrofits and new construction because of DLC's unrealistic minimum lumen and maximum CCT requirements. People can get rebates by retrofitting a 2x4 troffer with one high lumen 32W fluorescent F32T8 and a .71-1.00 fixed BF ballast, which provides 1760-2480 out of troffer lumens and a very good amount of light, especially if there are good task lights. But, an LED solution would require providing at least 3000 out-of-troffer lumens to be approved by the DLC and qualify for many rebate programs. The 3000 lumens could actually be too bright. The DLC also only allows up to 5000K, even though higher CCT LEDs are more efficient and can be better for alertness, based on human centric lighting.

California Title 24 takes effect on July 1, 2014. Prescriptive rebates will probably not be based on how much energy is saved compared to existing, but instead compared to the new worst allowable power densities, which are quite low. Other states and organizations may try to do something similar.

Chapter 26

Follow the Money

INTRODUCTION

Since the final step is the end customer, that is the perspective in this chapter. As you are probably already aware, following the money can be so important regarding almost every financial transaction, including lighting. It is the process of finding why someone is trying to promote or sell something. To be fair, the reason is not always bad. Please do not get me wrong, I think that many lighting people and firms are honorable, but if you get certain signals from people and firms, this chapter may help you clarify what those signals imply.

ONE-TRICK PONY

Sometimes one technology is the best choice, but not always. If dealing with a sales person with a company which offers only one technology (which could be LED, fluorescent, CMH, induction, etc.), what do you think that person is going to push? Almost all the time, it will be what that company offers, even if that is not the best for a specific project. Thus, it is often good to deal with a company that offers multiple technologies, because usually when their sales people offer recommendations, they will be more credible than those from a one-trick pony.

LIGHTING REPRESENTATIVE AGENCIES

- They probably will only recommend products from manufacturers whom they represent.

- What they call an equivalent may or may not be equivalent.

- Agencies want to package the entire project with their manufacturers.

- Agencies are often the biggest factor in pricing.

- Agencies often do not understand lighting retrofits, and most of the lighting retrofit products do not go through representative agents.

DISTRIBUTORS

Although distributors can usually get lamps, ballasts, fixtures, etc. from various manufacturers, they are often closely associated with specific manufacturers with whom they can get the best prices.

For example, for fluorescent lamps the connection is usually with GE, Philips, or Sylvania, and the distributor will probably push just one of them over the other manufacturers (even for specific lamps) when at least one of the others has a better product.

Also, several distributors offer free vacations and other benefits to contractors who buy a lot from them, which can motivate contractors to buy primarily from them.

CONTRACTORS

In addition to getting benefits from distributors, some contractors may get leads from a manufacturer such as one of the big three fluorescent lamp companies. As long as the contractors keep getting leads from the manufacturers, they will probably keep buying from those manufacturers. However, the products from a specific manufacturer may not be the best for a specific project. The question is, will the contractor use the best products, even though that may jeopardize getting leads in the future?

LIGHTING CONSULTANTS & DESIGNERS

They can get free dinners and "training sessions," which are mainly vacations in nice places with limited training. This practice can impact what they recommend.

Some consultants and designers consult for specific manufacturers. Good lighting professionals will volunteer full disclosure about being compensated by certain manufacturers for certain work. However, there are some lighting professionals who are so gung ho about certain technologies or manufacturers that they make me wish they had to take a lie detector test and be the subject of an investigation as to whether they are getting some kind of indirect or direct compensation.

STAN WALERCZYK

What about me? First of all, I will not accept any sales commissions. Having worked for decades to establish good credibility, I will not do anything to damage it.

I have consulted for several lighting manufacturers, which up through the beginning of 2014 are:

- A.L.P. Lighting
- Energy Solutions International
- Envirobrite
- Osram Sylvania
- Philips
- PlanLED

I have presented seminars for Cooper, GE, and Holophane to end customers, contractors, ESCOs, distributors, etc.

In my webinars and seminars for AEE, IES, various utilities, etc., I list specific manufacturers and models but clearly state that I am not endorsing anything. If I just say something like "high performance," people may not know what I mean. But if I state manufacturer A and model B, people can check out the lumens, LPW, etc. and know that any other product with the same or better specs can also be considered high performance.

All of my white papers have a "follow the money" section mentioning which company paid for at least most of my time to write it. Without that support, I could not afford to write white papers and provide them for free. Those white papers include other manufacturers of similar products, if there are any.

JOHN FETTERS

John is a principal of Effective Lighting Solutions and is a top-notch, credible lighting consultant.

He also presents for AEE. We co-teach Advanced Lighting Retrofit Options for AEE.

Chapter 27

Alphabet Soup

This is a short chapter, but it is an important one.

It is usually very good to get yourself some "alphabet soup," which means at least one good certification after your name, which can help getting a job, being promoted, and/or getting sales.

People understand that one must keep learning in order to renew certifications, which means such professionals keep up to date. I always recommend to end customers that they deal with lighting professionals who have least one good certification, and I tell them more than one certification is better. Following are some of these, in alphabetical order.

BOC
Building Operator Certification
Northwest Energy Efficiency Council's certification
Although this is not a dedicated lighting certification, lighting is included.

CALT, CSLT, CLMC & CSLC
Certified Apprentice Lighting Technician
Certified Senior Lighting Technician
Certified Lighting Management Consultant
Certified Sustainable Lighting Consultant
International Association of Lighting Management Companies' certifications

CEM
Certified Energy Manager
Association of Energy Engineers' certification

Although this is not a dedicated lighting certification, lighting is included.

CLEP
Certified Lighting Efficiency Professional
This is one of Association of Energy Engineers' certifications, which I have had for a long time.
For lighting retrofitting and maintenance, this is probably the best.

HCLP
Human Centric Lighting Professional
This is a new certification from the Human Centric Lighting Committee, which I also have.

LC
Lighting Certification
National Council on Qualifications for the Lighting Professions Certification
This is more for lighting designers; I also have this.

LEED AP
Leadership in Energy & Environmental Design Approved Professional
There are several versions of AP from the U.S. Green Building Council.

GENERAL
AEE has several other certifications, which could be researched.

Chapter 28

Big Brother Lighting

INTRODUCTION

Is "Big Brother," the Orwellian type, even with the best intentions, really helpful in lighting? Do organizations really know what is best for lighting vendors and end users on specific projects? Maybe they did when LEDs were really a new technology, but what about now? Why take choices away from lighting professionals and end customers, when it is they who usually know what is best for specific projects?

Some regulation is good, and the organizations that have provided good regulation should be commended because it can save energy and prevent some lighting vendors and end customers from engaging in bad practices. But, when unpractical regulations have been made, they should be questioned.

You can evaluate the current activities of several organizations. These include DesignLights Consortium (DLC), American Society of Heating, Refrigerating and Air-Conditioning Engineers (ASHRAE), California Energy Commission (CEC), and numerous rebate providers. This chapter was written on August 1, 2013; after that, some organizations may improve, stay about the same, or get worse.

Although this book is about lighting, these "big brother" issues also apply to building envelope, HVAC, motors, etc.

EXAMPLES OF BAD REQUIREMENTS

Following are just a few examples of bad requirements, especially for retrofits.

Some codes require occupancy or vacancy sensors, even though they may be just an added parts and labor cost and may increase burn time in many private offices and elementary classrooms where office workers, teachers, and/or energy police students have been doing a great job by manually turning off lights when they leave an area. Often, after occupancy sensors are installed, people allow the 12-15 minute delay before the sensors automatically turn off the lights. Even if the occupancy sensors reduce burn time, often the power density with a lighting retrofit is so low that paybacks and other financial returns are not very good for sensors and other controls.

Some codes require bi-level lighting. Yes, that may save energy in many rooms, but often it will not save any energy and just be an additional cost.

Rebates may not be available on LED troffers or troffer kits because those products do not meet impractical minimum lumen and maximum CCT requirements, even if the products do a great job in offices with good task lights, halls, restrooms, etc. Fluorescent equivalents do qualify for rebates.

Regarding retrofits and new construction, it is very bad that DLC is not allowing over 5000K for interior applications (at least until early 2014), although over 5000K can be so good for improved circadian rhythms and alertness.

RED FLAGS

Although there are attempts to change specific policies in specific organizations, and the various rebate providers may be very worthwhile, there is a bigger picture that includes the process that organizations use. At least some of the current processes may be similar to those of the U.S. Congress. (Think lobbyists and money.) So, any of the following should raise red flags.

Decision makers may include too many "ivory tower" experts and "big brother" advocates who do not have experience working in the trenches, which is what it takes for many energy efficient lighting retrofit projects to be approved and implemented.

Some key people may want requirements because of their ego and/or part of their legacy. Maybe one organization wants to look "better" than others with lower power densities.

Does it look like one or more key people are selling a state, region, or country to dimming and control manufacturers?

There may be numerous meetings open to the public, but it may be that, for the most part, only large manufacturers and/or manufacturers who want policy to favor the type of products they make can really afford the time and money to attend.

Paid and unpaid experts may be "hired guns" getting direct or indirect compensation from individual manufacturers or a manufacturers' association. It might be very good for organizations to do thorough background checks to see if experts have or will receive any compensation from manufacturers or related associations, and to require experts to sign a document pledging that they will not do so. Even if manufacturers or related associations do not compensate them, some of these experts may want something like stringent dimming and controls requirements implemented, because that may help them get more design work with end customers.

Lastly, do you think that at least some of your rights as a lighting professional or end user are being taken away with regard to determining the best lighting, with or without controls? Please do not get me wrong; various organizations and rebate providers have done a lot of beneficial work. But, are they too much "big brother"? too fond of stick rather than carrot? too prone to favor some technologies over others without good reasons?

Chapter 29

Human Centric Lighting

INTRODUCTION

This is the last chapter because, if it were any earlier, you might not have read anything after it.

WHAT IS HUMAN CENTRIC LIGHTING?

Figure 29-1 provides some clues. Human centric lighting (HCL) can also be called human factors in lighting, biophilia, and other terms. It is lighting for human health and wellbeing. HCL includes daylight and the new generation of tunable (dimming and warm to cool white or full range of color changing) LED products. Some tunable LED products can also change color saturation and proximity to a black body curve. Basic fixed output and fixed-Kelvin lighting products can often work very good at specific times of day, such as with 5000K and higher CCT products at work and school, and lower light level and warmer color tone products 1-3 hours before going to bed. I firmly believe that HCL will be the next big step in lighting, perhaps becoming even more significant than Edison's creating the light bulb.

Circadian Rhythms

Parts of this section are based on the work of Dr. Steve Lockley at Harvard School of Medicine and Dr. George Brainard at Thomas Jefferson University, as well as others. Dr. Lockley recently wrote a book, *Sleep: A Short Introduction*, which may cost less than $11 at Amazon or other venues. You may find of interest the Dr. Brainard video at www.youtube.com/watch?v=Pwg8s4B_

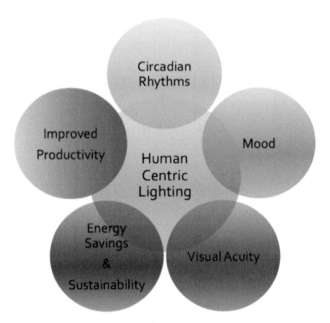

Figure 29-1. Human Centric Lighting

Cyw. Hopefully, it will still be available down the road.

Humans have been around for at least 150,000 years with the earth's natural lighting cycle. This cycle has a low light level and warm CCT in the early morning, a high light level and cool CCT during mid-day, a low light level and warm CCT again during the evening, and an extremely low light level and medium CCT on a moonlit night. There are short durations of very cool CCT near sunrise and sunset, but the light levels are very low. We developed a 24-hour circadian rhythm, or internal clock, under these conditions. The recently discovered intrinsically photosensitive retinal ganglion cells (ipRGC) are very important in setting our internal clocks. They are especially responsive to light that is rich in blue content (like the mid-day sky), which can be up to 10,000K and even higher.

Until 200 years ago, 90% of our waking time was spent outside. Now most of us spend 90% of our time indoors with electric lighting. While we're at work, our lighting is usually set at one light level with a constant CCT, which is not consistent with how our internal 24-hour clocks developed. Figure 29-2 shows two daylight spectral distribution charts that show the lighting we evolved with.

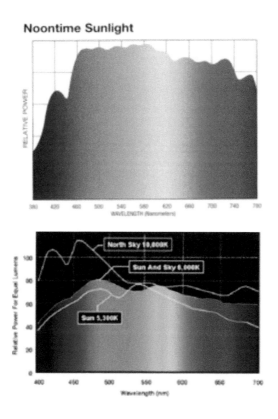

Figure 29-2

Even without clouds, there are numerous changes, due to air pockets and wind, which may be important to humans. Dieter Lang at Osram in Germany provided the daylight chart shown in Figure 29-3. ("Zeit" is time.)

Light and darkness control hormone production. With a proper circadian rhythm, there is a sufficient amount of hormones as follows.

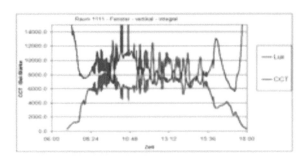

Figure 29-3. Daylight chart

During the day:
- Dopamine —for pleasure, alertness, and muscle coordination
- Serotonin—for impulse control and carbohydrate cravings
- Cortisol—for stress response

During the night:
- Melatonin—for allowing sleep and refreshing our body

Blue light content suppresses melatonin and encourages dopamine, serotonin, and cortisol production, resulting in people being more alert and productive at work, during day or night shifts. Our breathing rate is faster, body temperature higher, and heart rate faster during the day, with the blue rich light. Light in the 460-500 nanometer range, with is blue rich, is generally considered the most important for this. The bottom line is that with proper circadian rhythm, people are more alert during the day and sleep soundly at night, which benefits work, school, healing, and just living well.

The Lighting Research Center developed Table 29-1 for the Department of Energy (DOE), showing illuminance for predicted 50% melatonin suppression, assuming a one-hour exposure during the early part of the night and pupil diameter of 2.3 mm. Some of the numbers are controversial; not all researchers agree with them. For example, why does it take more light from 4100K fluorescent lighting than from 3350 K fluorescent when the 4100K lighting has more blue content? Also, why is there about the same light level with 5200K LED and 3350K fluorescent, when the 5200K LED has considerably more blue content? It is probably safe to state that generally higher CCT lighting suppresses more melatonin than lower CCT lighting.

Table 29-1. Illuminance for predicted 50% melatonin suppression

Light source	Illuminance (lx)
Daylight (CDL D65)	270
2856 K incandescent A-lamp	511
2700 K CFL (Greenlite 15WELS-M)	722
3350 K linear fluorcscent (GE F32T8 SP35)	501
4100 K linear fluorescent (GE F32T8 SP41)	708
5200 K LED phosphor white (Luxeon Star)	515
8000 K Lumilux Skywhite fluorescent (OSI)	266
Blue LED (Luxeon Rebel λ_{peak} = 470 nm)	30

Just using a Kelvin metric may not be sufficient with electric lighting. There is also the non-vision component of sight, the connection from the eye to other parts of the brain. The brain can fill in the gaps between the three spikes for vision in tri-phosphor fluorescent lamps, but these lamps may provide too little or too much light at certain wavelengths.

For example, look at the difference in wavelengths between two GE 5000K T8 lamps as presented in Figures 29-4 and 29-5. The Chroma is a full spectrum lamp (Figure 35), and the SPX is a tri-phosphor lamp (Figure 29-4). Full spectrum fluorescent lamps are not used much because of low lumens and high cost, but they have more overall energy in the 460-500 nm range. (Please be aware that the radiant power benchmark numbers on the vertical grids of Figures 29-4 and 29-5 are different.)

Note the representation of wavelengths for the GE SPX30 T8 lamp (Figure 29-6) and the GE SPX41 T8 (Figure 29-7). You can see that the 3000K lamp (Figure 29-5) has much less energy in the 460-500 nm range (good for the evening before going to bed but not good during the day, when you want to be alert).

It gets very interesting when analyzing 3000K tri-phosphor

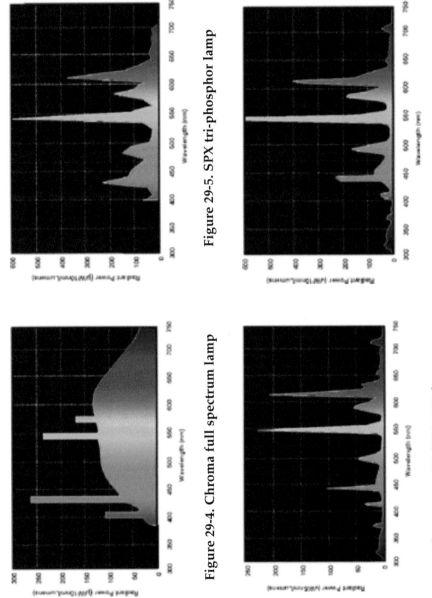

Figure 29-4. Chroma full spectrum lamp

Figure 29-5. SPX tri-phosphor lamp

Figure 29-6. GE SPX30 T8 lamp

Figure 29-7. GE SPX41 T8 lamp

lamps and 3000K LEDs. These warm fluorescent lamps have very little content in the 460-500 nanometer range, which can be good for lighting in the evening before sleeping. But, white light LEDs are really blue LEDs with a phosphor similar to what is used in fluorescent lamps. While it seems that most 3000K LED do not have that much energy in the 460-500 nm range, you should get spectral distribution charts for ones that you may want to use in the evening before going to bed, to be sure the blue content is low.

The white LED vs. incandescent chart (Figure 29-8) was also provided by Dieter Lang. The LED has the two bumps, and the incandescent is the straighter line. There does not seem to be much difference between the two in the total energy at 460-500 nm. But, at certain nanometers within that range, each has more or less energy, which may become more important after science determines peak sensitivity for melatonin and other substances within that range. You can also compare this chart with the fluorescent spectral distributions (Figure 29-2).

One way to reduce energy in the 460-500nm range is to use Tri-color RGB LEDs or Multi-emitter LEDs, which are RGB LEDs with at least one more colored LED (RGB+). These LEDs still provide a good white light. Compare the spectral power distribution charts in Figure 29-9 with those of typical phosphor white LEDs

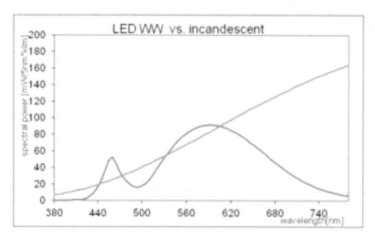

Figure 29-8

and tungsten incandescent.

Although some lighting professionals consider the light from incandescent bulbs the "holy grail" and similar to daylight, you can compare the daylight and incandescent spectral distribution curves and see how different they are. Incandescent bulbs have very low energy at the low end of the spectrum.

In 2012 the AMA adopted a policy recognizing that exposure to excessive light at night can disrupt sleep, exacerbate sleep disorders, and cause unsafe driving conditions. The policy also supports the need for developing lighting technologies that minimize circadian disruption, and it encourages further research on the risks and benefits of occupational and environmental exposure to light at night. AMA board member Alexander Ding, M.D. has stated:

The natural 24-hour cycle of light and dark helps maintain alignment of circadian biological rhythms along with basic processes that help our bodies to function normally. Excessive exposure to nighttime lighting disrupts these essential processes and can create potentially harmful health effects and hazardous situations. This type of disruption especially impacts those employed by industries requiring a 24-hour workforce as well those faced with unsafe driving conditions caused by artificial lights on cars and roadway illumination. By supporting new technologies that will reduce glare and minimize circadian disruption, the AMA is taking steps to improve both public health and public safety.

Philips has developed a lighting system called HealWell. This system uses high-lumen, high-CCT lighting for morning to mid-afternoon, and then lower-lumen, lower-CCT lighting for late afternoon. Results from at least one European hospital show that patients go to sleep earlier, sleep longer, heal faster, and leave the hospital earlier. There was also improved patient satisfaction and support recovery in patient rooms. Be aware that this, like several other studies from manufacturers, has not, to my understanding, been peer reviewed. One of several videos and write-ups on this can be found at: www.newscenter.philips.com/main/standard/news/press/2011/20111122-healwell.wpd The collage of one room in Figure 29-10 shows high CCT and light levels on

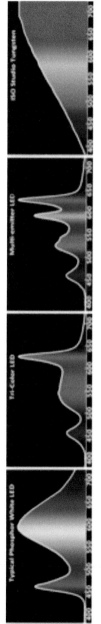

Figure 29-9. Spectral power distribution charts

Figure 29-10. Example of varying light levels

the left, representing morning; mid CCT and light levels in the middle, representing late afternoon; and low CCT and light levels in the evening. A video showing different light levels and CCTs in a patient bedroom throughout a complete day can be found at: www.lighting.philips.co.uk/application_areas/healthcare/healwell.wpd If that does not work, Google search "Philips Heal-Well."

Philips has completed case studies with high and low CCT using fluorescent lamps and dimming electronic ballasts, which can provide a full range of very warm white-to-white light rich in blue content. (The company is transitioning from fluorescent to tunable LED.) This type of lighting is also good for office workers, students, and others, including elderly people.

"The Effect of High Correlated Colour Temperature Office Lighting on Employee and Work Performance" was published in the *Journal of Circadian Rhythms* in 2007. The article presented a case in which fluorescent lighting with 17,000K lamps was in installed in an English office building. The conclusion was: "High Correlated colour temperature fluorescent lights could provide a useful intervention to improve wellbeing and productivity in the corporate setting, although further work is necessary in quantifying the magnitude of likely benefits."

Everyday of the year, the dimming and Kelvin-shifting system could be controlled in an office, hospital, home, etc. to match the outside CCT throughout most of the day, which may help keep proper internal clock alignment.

Some people have a hard time going to sleep at night. One culprit may be computer monitors and other electronic displays that are frequently used before going to bed. Most of these devices are relatively bright and have a high content of blue light. One thing that may help is a software program called flux, which makes the color or your display adapt to the time of the day—cool during the day and warm at night. There is a free download at: http://stereopsis.com/flux/

Alarm clocks with blue light may also be a problem. Ones with red light would be better. Some hospital ICU halls and nurse stations have red light at night.

There is also jet lag, which shifts your normal day/night either forward or backward. It may be good to have more blue light later at night when flying west and more blue light earlier in the morning when flying east. Boeing is using tunable LED lighting in the 787 Dreamliner cabins. (See Figure 29-11.) But, some of the programmed color tones may not be best at certain times of the night. The company may make the proper adjustments.

Figure 29-11. Airplane tunable LED lighting

Such an accomplishment was especially important for the San Francisco Giants, the west coast baseball team that won the World Series in 2010 and 2012. Lawrence D. Recht, Robert A. Lew, and William J. Schwartz, in Nature (Vol. 377, October 19, 1995) stated, "Whether due to jet lag or some other etiology, our findings are of practical importance for the west coast teams, because they face the double handicap of playing away games after eastward trips. The result is that these teams have been giving up more than one additional run in every game played after such travel."

Lighting is very important as we age. Even at 50, people need more visible light due to yellowing of the eye lenses. Sometimes the amount of light should be twice as much as that for teenagers and people in the their early 20s. Also, there should be no blue rich light in the evening.

Often Alzheimer patients have short periods of sleep and wakefulness, both day and night, making it even harder on caregivers. There have been some studies using rich blue light in the morning and afternoon, which may help such patients stay awake during the daytime and sleep during the night. This can be a big help to caregivers. The Lighting Research Center has done considerable research on this.

Mood

As indicated before, the hormones related for circadian rhythms affect mood during the day, including dopamine for pleasure, serotonin for impulse control and carbohydrate cravings, and cortisol for stress response.

Other lighting factors can also improve mood. When you walk into a room, you can get a feeling—positive, neutral, or negative. A positive feeling can improve mood. Parabolic troffers, which can create the dreaded cave effect, can cause a negative feeling and mood.

Giving people options can improve mood. For example, many office workers, especially those in cubicles, do not have very much control of their space. Being able to turn LED tunable task lights on and off as well as aim, dim, and adjust color tone could improve a worker's sense of control, mood, satisfaction, and productivity.

When reclining in a dentist chair or lying on a doctor's exam table, would you rather look up at a typical troffer with prismatic lens or at a troffer that has a lens with a blue sky and some clouds or other nature scene? Although several companies have been using nature scene lenses with fluorescent lamps, much of the fluorescent light does not get through the lens; therefore, a relatively high wattage is necessary. LEDs, which are directional, can much better penetrate these nature scene lenses.

In 2012 I visited the lobby of Virginia Mason Hospital in Seattle. The PlanLED sky panel troffers (see Figure 29-12) improved the ambiance of the space, where people can tend to be nervous while checking into the hospital. See: http://www.youtube. com/watch?v=rYDW7rvbpTY&feature=channel.

Figure 29-12. PlanLED sky panel troffers

Fraunhofer's virtual sky system (see Figure 29-13) has programmable RGB LEDs, allowing for numerous effects, such as clouds passing over the ceiling. Although the pricing was about $1000 per square yard in 2012, that should come down dramatically and become more cost effective in the future. See: www.fastcoexist.com/1679095/can-natural-light-make-employees-more-productive.

Figure 29-13. Virtual Sky System

Research conducted by Hella and Hamburg University in Europe has indicated that LED designs can significantly affect the comfort level of drivers and passengers. Different colors have been found to provide a wide range of emotional responses, from excitement to feelings of calm. (See LEDs Magazine, June 26, 2012.)

Now, let's discuss seasonal affective disorder (SAD). It is generally treated as a form of depression, affecting an estimated 10 million or more people in the United States, according to the American Psychological Association. Symptoms include changes in mood, appetite, and sleeping patterns during the fall and winter months, when natural light is typically at its lowest levels.

SAD seems to be considered much more important in Europe than in the States. Figure 29-14 shows a map from the Weather Channel in England. Why don't we have this in America? Light with a lot of blue content, especially in the morning, can significantly help with mood. Often this is done intermittently.

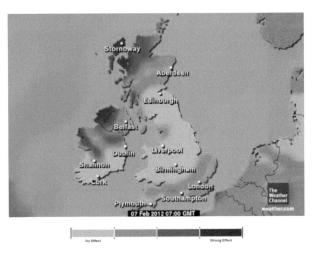

Figure 29-14. Weather Channel map (England)

Traditionally white light providing as much as 1000 foot-candles has been used. But even with fluorescent or LED, this consumes a lot of energy. (See Figure 29-15.)

There are now some blue LED lights with much less light output and wattage, but it is important to get those without too much infrared light, which may cause macular degeneration and/or other diseases. (See Figure 29-16.)

Light lounges can be built to include both daylight and elec-

Figure 29-15

tric light to reduce SAD. Although this is much more common in Europe, the University of Northern Iowa remodeled part of its Sabin Hall to be a light lounge. During the day, there is sufficient light from the sun; during evening, high light level electric lights are used, as shown on the right in Figure 29-17. Mike Lambert at KCL Engineering was the senior lighting designer.

Again, there is so much interrelationship. Better mood may improve productivity, performance, sleep, and other parts of our lives.

Figure 29-16. Blue LED light

Visual Acuity

Although there is a separate chapter on spectrally enhanced lighting (6), which is so important, it is also described here to help the flow of this chapter.

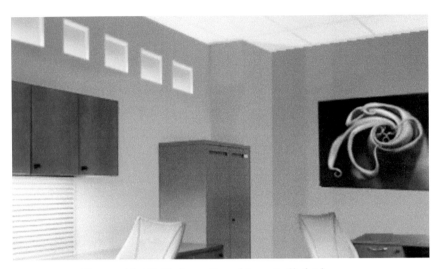

Figure 29-17. University of Iowa's light lounge

Much of this information is based on the work of two gentle-men. Dr. Sam Berman developed spectrally enhanced lighting, formerly known as scotopically enhanced lighting. He is now Se-nior Scientist Emeritus at Lawrence Berkley National Laboratory. Brian Liebel has been directing the DOE research on spectrally enhanced lighting and (at least through early 2013) is the chair of the IES Visual Effects of Lamp Spectral Distribution Committee, which wrote TM-24. They worked together on these and other projects over the years. I assisted on the DOE research and have been a member of this committee.

This DOE website may be very helpful. If the address does not work, you could Google search "DOE spectrally enhanced lighting." Several of the case studies are mine. See: http://www1.eere.energy.gov/femp/technologies/eut_spectral_lighting.html.

The IES Board approved TM-24 and an accompanying position statement. One key tool in TM-24 is equivalent visual efficiency (EVE), which shows how much less wattage can be used with higher-Kelvin lamps compared to lower-Kelvin lamps, while providing equal visual acuity.

Remember the ipRGC photoreceptors for circadian rhythms? They are also very important for visual acuity.

Visual acuity is affected by the color of light. Light sources that have relatively higher amounts of blue light stimulate the ipRGC photoreceptors, which in turn makes the pupils of the eye smaller, and this results in better visual acuity so a person can see more clearly (under otherwise identical lighting conditions and measured photopic foot-candle levels). The correlation of lamp spectral distribution and pupil size can be described by the scotopic/photopic (S/P) ratio of the lamp; higher S/P values indicate more blue in the light spectrum.

In general, the relative amount of blue in a light source is also correlated with the CCT rating, although there are variations among lamp types. An example benefit of this is being able to read one smaller row of letters on a Snellen eye chart with 5000K, compared to 3500K fluorescent lighting under the same photopic foot-candles.

The term for using spectrum in lighting design to affect visual acuity is spectrally enhanced lighting. Spectrally enhanced lighting can be used mainly in two ways. One is to maintain existing wattage and improve visual acuity. The other is to reduce wattage and maintain visual acuity. Most of the time, the second strategy is used.

Over the last 15 years, about 90% of my retrofit projects have been with 5000K or higher fluorescent lamps, with the goal to save maximum energy while still providing good lighting.

John Fetters, at Effective Lighting Solutions, who has co-presented many webinars through AEE with me, has specified, at 5000K or higher CCT, the fluorescent lighting for many military and other retrofit projects. Numerous other lighting professionals have done the same. Many lighting contractors, after having initial doubts about the merits of spectrally enhanced lighting, have changed to the 5000K lamps as standard once they have personally experienced the installations and benefits of occupant satisfaction and energy savings.

An important question concerns the main purpose of a

project. Is it for spectrally enhanced lighting, which can save considerable energy, or is it mainly for improved circadian rhythms, which can lead to improved worker productivity? There is considerable overlap. For example, 5000K has considerably more blue content than 3500K, so people may be able to be more alert (but perhaps not as much as with 6500K).

For just spectrally enhanced lighting and energy wattage reduction, 5000K may be the best choice because:

- The color is not too blue.
- Little or no loss of photopic lumens.
- Fluorescent lamp costs are not higher.

For improved circadian rhythms during work or school time, you can use 6500K, 8000K, or even higher. However, this can look too bluish to some people. Other considerations are:

- Even with the higher S/P ratio and a decrease in photopic lumens, there's not that much gain over 5000K in regard to the benefit of spectrally enhanced lighting.

- Fluorescent lamp pricing will be higher.

Following are two of Rod Heller's projects with 8000K fluorescent lamps, which provide both spectrally enhanced and circadian rhythms benefits. Rod manages Energy Performance Lighting.

In a computer intensive environment at American Family Insurance, ambient light levels were lowered to 15 foot-candles using 8000K lamps. Common responses from employees have been: "It is easier to see," "I am not tired at the end of the day," and "I sleep better at night." This was not a scientific study, but the initial reactions to these lamps have been very positive, and we suspect research will bear out similar results.

At Veyance Technologies, which manufactures brake hoses for automobiles, 5000K lamps were installed in the production area 5 years ago. The company sampled 8000K lamps in a small area of the plant. The employee response to the 8000K lamps was so positive that the entire manufacturing plant was relamped with 8000K lamps. The reasons given included: "I feel better," "It is easier to see defects" (critical in his company), and "I sleep better." The general manager of the facility stated, "This is the best light I have ever worked under and I cannot understand why everybody does not work under it." Even the ISO 9000 inspector gave them high marks for their lighting based on visual acuity.

Although Rod has seen and heard claims that people should not have 8000K lighting for more than 3-4 hours per day, nor after about 1 PM, neither he nor I have seen any good scientific research supporting those claims. If anybody knows of any good scientific reports backing up those claims, please let me know. Common sense indicates to me that generation after generation of human beings have been outside for long periods of the day with up to 10,000K.

Many lighting designers and architects do not generally like anything higher than 3500K with fluorescent or LED, and many consider incandescent, which is about 2700K, the holy grail even though there is way more red and way less blue and green than daylight most of the day.

Figure 29-18 represents another spectral distribution chart for incandescent, which you can compare with charts from various Kelvin temperature fluorescent and LED sources.

I find it quite interesting that some lighting designers who promote daylighting, which is typically a high CCT most of the day, want low-Kelvin electric lights in those same commercial spaces. Even with daylight harvesting, the electric lighting usually still needs to be on to some degree for portions of the day. Those lighting designers and architects may give their clients three initial choices:

- 3000K, which is like warm white T12
- 3500K
- 4100K, which is like cool white T12

Then, when people pick 3500K, those lighting professionals feel con-

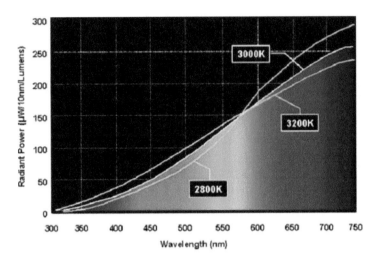

Figure 29-18. Spectral distribution chart for incandescent

fident that they are correct, thinking that 3500K is neutral and best.

It has been my and some other people's experience that if you give people three initial lighting choices, they usually pick the one in the middle, as indicated above. Thus, if you give people 3500K, 4100K, and 5000K, they usually pick 4100K. If you give them 4100K, 5000K, and 6500K, they usually pick 5000K, etc.

Actually, initial preferences are not that important. Even if you give people the best solution since sliced bread, they may not like it initially because they have been used to something else for a long time. It can take them two to three weeks to make the internal adjustment for the new lighting. In the first rounds of some DOE research (for office buildings with real office workers), although a number of people did not like the 5000K lighting initially, many of them became very strong supporters of 5000K after three weeks. In general, there was no statistical preference for lower Kelvin compared to 5000K after three weeks.

Some people have attributed the results of this DOE research to the Hawthorne Effect, which is not an accurate view. (As many are aware, the Hawthorne Effect describes a situation where worker activity improves simply because the workers are singled out and made to feel special.)

Some people try 5000K or other high-Kelvin lighting without delamping or using a lower BF ballast, then complain that it is too bright. They prove the point that high Kelvin lighting works, but they really need to delamp or use a lower BF.

Over the years, a lot of offices, schools, and other commercial buildings were originally constructed using 3000K or 3500K fluorescent, but most retrofit upgrades use 4100K, 5000K, or even higher with good user satisfaction. For well over a decade many retrofitters, ESCOs, and I have been using 5000K, 6500K, and even 8000K. Although there have been limited projects in the world (some recently) with 12,000, 14,000, and even 17,000K, time and more peer-reviewed projects will be necessary to determine how really good such very high CCT lighting is. But, compared to 12,000 and higher Kelvin, 5000K can seem like warm white.

Tables 29-2 and 29-3 may be helpful in putting information together.

Table 29-2

S/P Benefits of 5000K 3100 Lumen F32T8s					
lamp	mean photopic (catalog) lumens	S/P ratio	brightness $P(S/P)^{.5}$	paper $P(S/P)^{.78}$	computer $P(S/P)^{1.0}$
F34T12 CW	2300	1.50	2817	3156	3450
F34T12 WW	2350	1.00	2350	2350	2350
F32T8 730	2650	1.19	2891	3035	3154
F32T8 735	2650	1.30	3021	3252	3445
F32T8 741	2650	1.56	3310	3749	4134
F32T8 830 2nd	2800	1.29	3180	3415	3612
F32T8 835 2nd	2800	1.41	3325	3661	3948
F32T8 841 2nd	2800	1.62	3564	4079	4536
F32T8 830 3rd	2950	1.29	3351	3598	3806
F32T8 835 3rd	2950	1.41	3503	3857	4160
F32T8 841 3rd	2950	1.62	3755	4298	4779
F32T8 850 3rd	2950	1.95	4119	4966	5753
		CW	46%	57%	67%
		WW	75%	111%	145%
Increase of energy efficiency of 3000+-initial-photopic-lumen 850 3rd generation F32T8s when considering full field of view compared to		730	43%	64%	82%
		735	36%	53%	67%
		741	24%	32%	39%
		830 2nd	30%	45%	59%
		835 2nd	24%	36%	46%
		841 2nd	16%	22%	27%
		830 3rd	23%	38%	51%
		835 3rd	18%	29%	38%
		841 3rd	10%	16%	20%
notes : Lumens and S/P ratios can vary among lamps and manufacturers.					
Prepared by Stan Walerczyk, www.lightingwizards.com, 1/1/14 version					

Table 29-3

S/P Info for 32W F32T8s and 34W F34T12s					
lamp	mean photopic (catalog) lumens	S/P ratio	brightness $P(S/P)^5$	paper $P(S/P)^{.78}$	computer $P(S/P)^{1.0}$
F34T12 CW	2300	1.50	2817	3156	3450
F34T12 WW	2350	1.00	2350	2350	2350
F32T8 730	2650	1.19	2891	3035	3154
F32T8 735	2650	1.30	3021	3252	3445
F32T8 741	2650	1.56	3310	3749	4134
F32T8 830 2nd	2800	1.29	3180	3415	3612
F32T8 835 2nd	2800	1.41	3325	3661	3948
F32T8 841 2nd	2800	1.62	3564	4079	4536
F32T8 830 3rd	2950	1.29	3351	3598	3806
F32T8 835 3rd	2950	1.41	3503	3857	4160
F32T8 841 3rd	2950	1.62	3755	4298	4779
F32T8 850 3rd	2950	1.95	4119	4966	5753
F32T8 865	2750	2.20	4079	5087	6050
F32T8 880	2518	2.50	3981	5146	6295
notes : Lumens and S/P ratios can vary among lamps and manufacturers.					
Listed F32T8 865 is Sylvania XPS. Listed F32T8 880 is Sylvania Skywhite XP.					
Prepared by Stan Walerczyk, www.lightingwizards.com, 1/1/14 version					

A general rule of thumb on energy savings with more spectrally enhanced fluorescent lamps is that 5000k can usually save 10% compared to equivalent 4100K, and 20% compared to equivalent 3500K. Often that turns out to be one lower BF ballast type per step. For example, each of the following provides about the same amount of light for the way the human eye really sees:

- 2 F32T8 835 & 1.15 BF ballast (approximately 72W)
- 2 F32T8 841 & 1.00 BF ballast (approximately 64W)
- 2 F32T8 850 & .88 BF ballast (approximately 54W)

Here is another example:

- 2 F32T8 835 & .71 - .77 BF ballast (approximately 47-48W)
- 1 F32T8 850 & 1.15 BF ballast (approximately 38W)

The DOE may perform S/P ratio testing on LEDs, which will be good, because to many people 3500K LED often looks more like 4100K fluorescent than 3500K fluorescent. But, there are some very good applications for 2700-3000K such as offices, restaurants, hotels, museums, and residential, with mainly warm color floors, walls, and furniture.

The Kruithof Effect (or Kruithof Curve) should be discussed, because some lighting professionals often use it to specify warm color tone lighting for low light levels. But the follow-up studies that I am aware of have not done a great job replicating the effect. You can judge for yourself. (See Figure 29-19.)

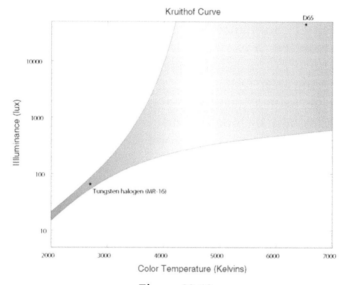

Figure 29-19

Brian Liebel came up with a very good point to ponder. There are many theories as to why different Kelvin lamps are popular in different parts of the world. One theory is that people near the equator prefer higher Kelvin, because the sun is typically higher in the sky there, thus appearing whiter. Brian came up with a new theory. Many people in the United States, Canada, and much of Europe got the incandescent bulb quite early and got used to that low-Kelvin light source. However, many people in many places around the world did not get electricity until after fluorescents became widely available, and those fluorescent lamps were typically high Kelvin, so the people got used to that.

The highest selling Kelvin lamp in the world is 6500K; it is practically the standard in China, India, Australia, Japan, Sub-Africa, and parts of the Middle East.

Looking down the road, tunable LED products ranging from 2700K up to 8000K or higher should become the best solution, because people will be able to pick whatever light level and CCT they want without having to replace lamps, like they have to do now with fluorescent.

Productivity/Performance

For working, it can be called improved productivity, and for learning, it can be called improved performance. For both, it includes being alert, feeling good, and being able to see well. The British study with 17,000K fluorescent lamps, discussed in the mood section, also included productivity improvements.

Philips developed a classroom lighting solution system called SchoolVision. This technique has been implemented in several European and at least one American school. With it, 12,000k is used during the first 30 minutes in the morning to shut down the children's sleep cycle and turn on their day cycle. Then 5000-6500K is used during normal study and learning activities, and the light level can be doubled for literacy instructions. After recess and during rest time, 2700K is used for a calming effect. Results reported so far have demonstrated increased academic performance, but peer review would be good. European and American videos are available at:

www.youtube.com/watch?v=3lfc1y8q5l4&feature=related
www.wtva.com/content/mediacenter/default.
aspx?videoId=6673@wtva.web.entriq.net&navCatId=17

The SchoolVision studies were done with various CCT fluorescent lamps in each fixture, with dimming ballasts, which works well; Philips will continue to improve them and will also develop LED systems. However, getting up to 12,000K LEDs with good color rendering and efficiency is currently a challenge.

As opposed to fluorescent, LEDs can maintain or even increase lumens per watt when dimmed, as long as the drivers maintain a good power factor when dimmed. Osram, the parent company of Sylvania, has performed lighting research studies in European schools, mainly with suspended direct/indirect LED fixtures. The direct light is 4000K, and the indirect light can be adjusted from 6500K to 14,000K. During the studies, the indirect light was kept at 14,000K. The spatial, spectral, and temporal properties are like outside, with higher CCT and brighter-looking horizontal and up to 45 degrees above horizontal. (See Figure 29-20.)

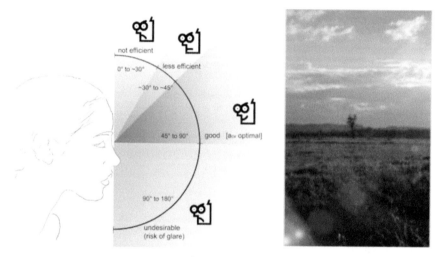

Figure 29-20

Dr. Katrin Hille, head of research at the Transfer Centre for Neuroscience and Learning, was in charge of the light study. Osram's press release, titled "New Light Creates Bright Sparks," included the following:

> Even though it has been recognised for a long time that light with a specific colour temperature and illumination intensity has a positive effect on people's performance and well-being, the fact that the students with the biologically optimised lighting had up to a third fewer errors in the concentration

test is impressive... Contributing to the positive results also was the fact that biologically optimized light provides stimulation to the body as if the person was outdoors. That is, the students' circadian rhythms shift forward and the young people became fit earlier. Therefore, this light can counteract social jetlag—tiredness in the mornings frequently observed in young people particularly. Many participants in the study are enthusiastic about the new lighting. "My concentration is actually better. In normal light, however, I sometimes have to keep myself awake in the classroom," as one student described the experience.

See Figure 29-21.

Figure 29-21

What is very important here is monetizing soft savings, which can be identified as improved worker productivity, student performance, or retail sales. For example, a quite conservative estimate is that better lighting can prompt office workers to waste five minutes less per day. That is 1% of an eight-hour day. If someone is making $50,000 per year, 1% of that is $500 per worker, year after year. Human resource people usually understand these benefits and can explain them to others in the organization. That totally dwarfs hard savings, which are mainly electrical savings, along with perhaps replacement parts and maintenance savings. Soft savings also dwarf replacement parts, labor costs, and rebates. (The same soft savings concept holds true with students scoring better on standardized tests, as well as with soft savings in retail sales.)

An incredible document to help decision makers understand the real value of soft savings is *The Economics of Biophilia—Why Designing with Nature in Mind Makes Financial Sense*. Terrapin Bright Green published it in 2012, and it is available on the web for free: http://www.terrapinbrightgreen.com.

These three points are in the abstract.

- Biophilia, the innate human attraction to nature, is a concept that has been recognized for several decades by the scientific and design communities, and intuitively for hundreds of years by the population at large.

- Today productivity costs are 112 times greater than energy costs in the workplace. We believe that incorporating nature into the built environment is not just a luxury but a sound economic investment in health and productivity, based on well-researched neurological and physiological evidence.

- Integrating quality daylighting schemes into an office space can save over $2,000 per employee per year in office costs, and over $93 million could be saved annually in healthcare costs as a result of providing patients with views to nature. (There is more in the Linking Desire for Nature with Dollars section.)

For this paper, indicators of productivity include the following and will be translated into dollars where most applicable:

— Illness and absenteeism
— Staff retention
— Job performance (mental stress/fatigue)
— Healing rates
— Classroom learning rates
— Retail sales
— Violence statistics

For anyone not aware, it can cost about half an annual salary to find and train somebody to replace an employee who wants to leave. Human resource people know this kind of information.

From the Driving Profit Margins in the Workplace section:

• In the last decade, American psychologists have aggregated the five strongest requirements for basic functioning that, if neglected, can trigger worker comprehension problems and dissatisfaction in the office space (Kellert, 2008). These are:
  — Need for change (varying temperature, air, light, etc.)
  — Ability to act on the environment and see the effects
  — Meaningful stimuli (Stagnant atmospheres cause an onset of chronic stress.)
  — One's own territory to provide safety, an identity, and protection
  — View to the outside world

• In 2010, the US Department of Labor reported an annual absenteeism rate of 3% per employee, or 62.4 hours per year per employee lost, in the private sector. Therefore, an employer will lose $2,074 per year to employee absences.

• Presenteeism describes the phenomenon in which workers clock in for work but are mentally removed from the workplace, causing labor-related financial losses for the company. Presenteeism can result from sleepiness, headaches, colds, and asthmatic drain if the air supply is poor. Presenteeism costs employers in the private sector $938 and employers in the public sector $1,250 per employee per year.

- From the US Department of Labor 2010, BLS 2011, BOMA 2010:
  — 86.3% Productive Salaries & Benefits
  — 0.8% Energy
  — 8.9% Rent
  — 2.7% Absenteeism
  — 1.3% Presenteeism

Electrical Savings & Sustainability

High-Kelvin fluorescent and LED ambient lighting systems can save a lot of energy. With lower energy there is also a lower carbon footprint. Watts per square foot (WSF) can be very low with tunable LED troffers, especially when there are also similar task lights. Offices can often have as low as 0.4 WSF, including both ambient and task lights.

Some of the tunable LED products cost about the same as their dimmable and fixed-Kelvin equivalents, but there is an additional cost for local controls and labor. With such low power density and the accompanying lower electric bills, complex and building-wide controls are often not cost effective.

With such a low carbon footprint and no mercury content, LED systems are quite sustainable. But, the DOE has not yet determined if LEDs are more environmentally friendly from cradle to cradle than incumbent lighting technologies. A 2013 DOE study showed that screw-in LED and CFLs were about even, including numerous factors. That study also showed that LEDs will become better than CFLs in the future.

Maybe better lighting will help workers themselves to be more "sustainable," taking less medically related time off. As a result, perhaps insurance premiums would be reduced.

Numerous corporations and organizations that have sustainability groups find that lower carbon footprint and less toxic chemicals can be more important than payback or other hard financial returns. This may be a good potential sales portal.

CONSIDER THIS

Let's say that you eat too much and come back to the office groggy, or you have to do some work late in the evening or during the night. Yes, you could drink coffee, cola, or another stimulating drink—or maybe you could increase the light level and CCT to stay alert without drugs. With evening or night work, keep in mind that you may be awake for an hour or so after you finish because of the drugs or the lighting.

IT WILL BE A TUNABLE LED WORLD, BUT THERE ARE SOME CHALLENGES

Other than daylighting, nothing is better for human centric lighting than tunable LED products. However, you will need to know more about the existing products and what to look for in upcoming products.

Two Ways to Change CCT

One way is to use RGB or RGB, plus an additional one or more other color LEDs, which can provide light that is quite close to daylight and all the variations of daylight. But, these systems are usually quite expensive and not very efficient. However, the evolution of using lime green LEDs compared to the existing standard green LEDs could make these systems considerably more efficient. Another concern with these systems is that since the various color LEDs have different lumen maintenance curves, there could be system color shifts down the road unless there is some kind of feedback control loop.

At this time, and maybe also in the future, the most cost effective solution is to use warm white and cool white LEDs together. An example is combining 3000K and 6500K, which are both white LEDs, and controlling them together or individually. In products like this, just the 3000K can be on for warm white lighting, and just the 6500K can be on for cool white lighting. Anything in between can be achieved by dimming one more than the other. Another example is using 2700K and 8000K white LEDs. In general, these products have the same LPW as fixed-Kelvin LED products. They also have about the same warranty.

The basic and least expensive version of the warm and cool white LED systems is using equal numbers of each color temperature and using individual zero to maximum dimming control. With a 3000K and 6500K system, the maximum light output would be with both fully on, and the Kelvin would be 4750. With just the 6500K LEDs fully on, photopic light levels would be half, but there may be sufficient light, based on spectrally enhanced lighting. With just the 3000K LEDs fully on, the photopic light levels would also be half. For late afternoon and evening use, it may be good to have a relatively low light level and warm white color tone, which can allow melatonin to be produced to help for good sleep. Since this version has the same number of LEDs as fixed-Kelvin LED products, the pricing can be about the same. So, with the same LPW, cost, and warranty, why not get tunable LED products instead of fixed Kelvin LED products?

An advanced tunable LED version allows the same maximum light levels at all CCTs, but to do this, more LEDs and a more complex control system are required, which increases cost. However, LPW and product warranty are still very good.

Flicker

Although most people thought that flicker went away with the transition of magnetic to electronic fluorescent ballasts, the flicker issue is coming back with many LED drivers, even at full light output and especially with deep dimming. There are two types of flicker—perceived and sensed. Perceived flicker means that people are aware of it. Sensed flicker is not consciously perceived. Both types can cause eyestrain, headaches, epileptic seizure, reduced visual task performance, stroboscopic effect, and distraction.

Flicker can be caused by the dimmer and/or the driver. The typical incandescent dimmer uses a phase-cut, or wave-chop, for parts of the AC sine wave. There are various types of AC LED dimmers, and some can cause more flicker than others. One common type is pulse width modulation, which, due to the increased time between the bars of power, may cause the LEDs to be off a greater percentage of time during deep dimming, resulting in flicker. When you want dimming LED systems, you should research dimmers and drivers. There are many LED systems with no or very little flicker at full output and deep dimming.

The DOE, through PNNL, has been working on how to measure flicker, etc. Hopefully, some publicly available information will be available by late 2013.

Dimming

Although numerous sales people, consultants, companies, and organizations push dimming fluorescent ballasts, they are not efficient. Dimming below .70 BF requires heating the lamp cathodes to prevent flicker, spiraling, and going off. This cathode heating consumes extra wattage. LEDs can get more efficient when dimmed because they run cooler. Depending on the driver, LPW levels can remain or improve while dimming.

Controls

Dimming controls do add cost, but if dimming is used, the additional cost to add Kelvin changing is minimal.

Various manufacturers are using wired as well as wireless control systems. Wireless versions include ZigBee, a TV remote control type, bluetooth, and wifi.

There are at least two ways to match daylight throughout most of the day. One is an exterior sensor; the other is a 365-day digital clock programmed for specific locations.

For open office areas, it is not recommended that each worker be allowed to dim and change the Kelvin of the ambient lights over his or her space, because some fixtures could be over more than one work station. With different color tones throughout the room, it could look ugly. All the ambient fixtures in a room should be set at a fixed color tone, or all should change together as the day progresses. But, each worker could have individual control of his or her tunable LED task lights.

In individual offices and other private spaces, the person could have full tunable control of ambient and task lighting.

INNATE SENSE

Although based only on very limited anecdotal evidence, I think that people have an innate sense of what light level and color tone should be used for various times of the day for various tasks. They do not need to be told—and I mean normal people, not lighting geeks.

In late 2012, I found it quite interesting to read reviews of people that purchased tunable LED task lights from Amazon. Based on reading those reviews, talking to people who have used these products, and my own experience, giving people the opportunity to dim and Kelvin-change their various lighting fixtures at work, school, and home at different times of the day for different tasks may really be the best thing since sliced bread.

TUNABLE LED PRODUCTS

The tunability of LEDs, even more than lumens per watt (LPW) and life, may be the biggest reason why LED will surpass high performance incumbent technologies. There is really no downside of many tunable LED products compared to fixed correlated color temperature (CCT) LED products. Numerous tunable LED products match the price, deliver up to 110 LPW, offer 100,000+ hour rated life, and may have up to the same ten-year warranty of standard LED products. If you are going to dim, the added price for controlling Kelvin or color changing is quite small. So why even consider fixed CCT LED products in many applications? Most of the early pioneers, such as Lumenetix, Osram Sylvania, Philips, PlanLED, and Telelumen have continued to improve their tunable products and reduce their pricing.

Task Lights

There are numerous manufacturers and models available, but check to see if they are Energy Star approved. In 2013, my favorite is the PlanLED TL-7000 because of these factors:

- Can provide 75 foot-candles by itself
- 84 CRI
- 2700-6500K
- Three-level dimming and color tone setting
- 12.2W at full light output
- Low glare
- Minimum shadowing
- Tall enough so it can light both in front and back of computer monitors for good uniformity and low contrast ratios
- Tall enough so it can mimic undercabinet task light
- Long reach, which is good to position over blue prints, etc.
- 50 minute timer
- White or black
- 50,000-hour rated life

There is a video available, but if the web address does not work, you can search YouTube PlanLED TL-7000. See: http://www.youtube.com/watch?v=XMM66hdD168.

Well-designed, tunable desk-mounted LED task lights may be the easiest and best way to introduce people to human centric lighting. Almost every time I have shown somebody a good tunable desk-mount LED task light, they have wanted to keep it. (See Figure 29-22.)

Troffers and Troffer Kits

Envirobrite and PlanLED (and perhaps others) have cost effective, tunable, hard-wired LED troffer kits and troffers. These have already started giving high perfor-

Figure 29-22. Turnable desk-mount LED task light

mance fluorescent retrofits a run for their money. Other North American manufacturers are expected to develop tunable LED troffers and troffer kits soon.

Screw-in Lamps with Cell Phone Apps
In 2013, there are already the Philips Hue (see Figure 29-23), Bluetooth Bulb, LIFX, Rambus, and Ilumi tunable screw-in LED lamps with cell phone apps. One feature on at least some of these is that you can aim your cell phone camera at something, such as a fireplace, and program the lights to mimic that light. Later on, the prices should be lower, with more products available.

Figure 29-23. Phillips Hue

Recessed Cans or Track Heads
Lumenetix, Pathway, and USAI had tunable LED recessed can modules, recessed cans, and/or track head modules available in 2013. The Lumenetix Araya has up to 97 CRI and excellent R9; the company has been working to improve LPW. (See Figure 29-24.)

Drum Fixtures
Here is a story on the PlanLED tunable LED, with 2' diameter, low profile, and 5000+ lumen drum. (See Figure 29-25.) My wife, who is not a lighting professional, really likes this fixture, maybe more than I do. Over the kitchen island we were going to have a ceiling fan and two recessed cans installed. The morning before

Figure 29-24

the contractors were going to cut holes for them in the ceiling, we plugged this drum fixture into an extension cord and gave my wife the remote control to play with the light levels, CCT arrows, and the preset buttons for reading, watching TV, and the night light. Right away, she decided on the fixture in the middle of the island instead of the fan and the two recessed cans. This type of fixture, already quite popular in Japan and some other Asian Pacific Rim countries, could be become common in North America.

Figure 29-25. PlanLED tunable LED

Rotatable Suspended Indirect/Direct Fixtures

During the second half of 2013, I was designing the lighting in the dining room, which is also used for numerous other activities, at an aged care facility. It currently has 52, 2'x4' ceiling box fixtures, each with 3 F32T8s that make the space feel commercial, are glary, and consume 4320W. PlanLED has been developing pendant indirect/direct concepts with wireless controlled independent dimming and color changing up and down light. Indirect/direct lighting with no more than 30% downlight is very good for older people. Figure 29-26 shows specifications for one 18-inch diameter concept that provides up to 100 LPW. Future concepts may have the drivers and controllers in a shallow, round-cone ceiling plate compartment, which would allow the light panels to be much thinner. Although rotating the light panels may not be used on this project, that could be very useful on other projects. If someone wanted more downlight than uplight, the panels could be rotated 180 degrees; if someone wanted to highlight something on a wall, the panels could be tilted that way.

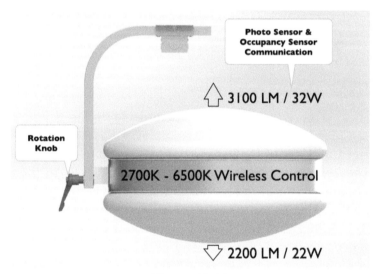

Figure 29-26

A Category Of Its Own

After my presentation about tunable LED products at the 2013 Lightfair, I walked the exhibitor aisles and was very impressed with the concept version of the Acuity Brands Winona Lighting Aera, which may be developed to replace overhead ambient lighting. The 6″ x 6″ recessed apertures are shown in the upper wall shown in Figure 29-27. Light levels, color, saturation, movement, etc. can easily be adjusted with Acuity's new Axion control system. An example is mimicking the view out of a high window at different times of the day, even with clouds passing from one side to another. Also, it can be set like an alarm clock. This may become available in 2014.

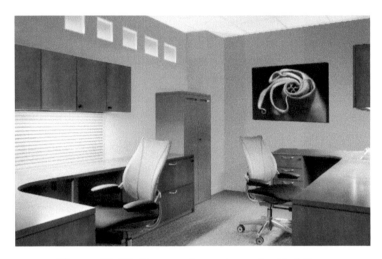

Figure 29-27. Recessed apertures for lighting

Another Category of Its Own

The Telelumen high fidelity system has up to 16 independent channels, 1 million levels of intensity, 1KHz response time, and 1000-100,000K. (See Figure 29-28.) Lumenscripts of a Maui sunset, moonlight hitting waves, noon daylight in Chicago, etc, can be produced and played at any time.

Exterior

Many existing LED exterior fixtures already dim, but new ones could be designed that may also change color tone. This might be good for circadian rhythms in both humans and animals.

FUTURE

What the future may really hold is described in two reports published in the Proceedings of the *Human Factors and Ergonomics Society's 56th annual meeting, 2012:*

Figure 29-28. Teleumen 2012 Replicator

Exploration of Gesture-Based Control For Tunable Solid State Lighting Applications was written by Jeremy M. Spaulding and Jeffrey Holt at Osram Sylvania. It describes how, just like the newest video games without hand-held devices work, lighting can be controlled by arm movements. Raising an arm up and down controls the light level. Side-to-side arm movement controls the CCT. Circular movements control color saturation.

A Closed-Loop Feedback System For a Context-Aware tunable Architectural Lighting Application was written by Jason Nawyn, Maria Thompson, Nancy Chen, Kent Larson, Massachusetts Institute of Technology, Media Laboratory, Osram Sylvania, Central Research, and Services Laboratory. This does not just include warm to cool white CCT changing but also colors such as green and blue in the periphery. Some colored LEDs may consume one-tenth the wattage of white LED lighting. Controls, including occupancy sensors and digital cameras, can help automatically adjust light levels, CCT, and color, depending the task. An example is using a lower light level for computer work than for a paper task.

Both excellent reports are available at: www.hfes.org//Publications.

HUMAN CENTRIC LIGHTING COMMITTEE

Since human centric lighting is so important, John Hwang of PlanLED, Rod Heller of Energy Performance Lighting, and I created the human centric lighting website. If you have not checked out this website and the useful information in it, please do at: http://humancentriclighting.com.

We also formed the Human Centric Lighting Committee. John Hwang, Mike Lambert, and I are the officers (in late 2013). PhDs include Dr. Steve Lockley at Harvard School of Medicine, Dr. Joan Roberts at Fordham University, Dr. Kevin Van Wymelenberg at the University of Idaho, Dr. Chris Johnson at University of Iowa and Dr. Michael Mott at University of Mississippi; Dr. Doug Steel at PhotoKinetics, Dr. Josep Carresas at the Catalonia Institute for Energy Research in Spain, and Dr. Marjukka Puolakka at Aalto University in Finland. Experts in certain fields include Eunice Noell-Waggoner, Rosalyn Cama, Brian Liebel, Jeremy Spaulding, Dieter Lang, and Milena Simeonova. There are also manufacturers, consultants, lighting designers, engineers, lighting retrofit contractors, energy center coordinators, utility representatives, rebate organizers, and others. There is considerable information, including questionable misinformation, and certain HCL Committee members can often inform people as to what is good, based on the best science.

Similar to the DOE working very hard to avoid the mistakes made with CFLs, we are developing a Human Centric Lighting Professional certification to help insure that human centric lighting projects are done well. Botched projects could give human centric lighting a black eye. We do not want to start from scratch, so people will need an LC, CLEP, AP, or another established lighting certification first.

We are also developing a Human Centric Lighting approved

product listing. This will require dimming with little or no flickering, Kelvin changing, at least 80 CRI, and 20 R9. At warm color tones, the amount of 460-500 nm light will have to be sufficiently low to help in falling asleep after using this lighting. Especially since the IES finally approved TM-24-13 (which is about the visual acuity benefits of high- Kelvin lighting), scotopic/photopic (S/P) ratios will be measured and listed. Again, we do not want to start from scratch. Most products will need to be already approved by Energy Star, DesignLights Consortium, or Lighting Design Lab. But, since at least the DesignLights Consortium (DLC) has ridiculous restrictions on above 5000K products, some products could get the Human Centric Lighting product approval without DLC approval.

Lastly, the Human Centric Lighting Committee would like to assist lighting designers, architects, engineers, contractors, ESCOs, and end customers in designing, installing, and commissioning as many good human centric lighting projects as possible, including hospitals, colleges, office buildings, retail stores, and other facilities around the world.

2013-2014 HUMAN CENTRIC LIGHTING PROJECTS

Following are some of the projects that I worked on in 2013 and some I will continue to work on in 2014.

Major League Baseball Team

Before home night games, the tunable LED lights in the home team locker room are set at a relatively high light level and a relatively high CCT to help the players become alert for the games. After the games, the light level is reduced and the CCT is a warmer color tone to help the players relax and perhaps go to sleep easier after they get home.

High Profile Office Building

Tunable LED troffers and task lights will be installed the in a section of this office builing as a mock-up, and since that went so well, much more of the building will be retrofitted later in 2014. This includes advanced controls for day-lighting, load tracking, etc.

Aged Care Facility

This was briefly discussed in the rotatable suspended indirect/direct fixture section earlier in this chapter. Hopefully, the new tunable LED lights will reduce accidents and improve alertness during the day and sleep during the night.

Elementary School

This project with skylights and tunable LED lights will be installed in the summer of 2014.

24-hour Call Center

This will include replacing parabolic troffers with top of office module wall LED uplights, tunable desk and under cabinet task lights, and SAD lights.

HUMAN CENTRIC LIGHTING PROJECTS LATER IN 2014 AND BEYOND

With your help, there should be many more human centric projects later in 2014 and afterwards.

LIGHTING IS NOT A COMMODITY

Especially after reading this chapter, I hope you consider lighting much more than just a commodity. There is much more to good lighting than just low wattage and proper light levels.

Thus, payback and ROI based just on electrical savings, rebates, installed costs, and sometimes maintenance savings should no longer be used. Soft savings, which include improved circadian rhythms, mood, performance, etc. should be included into the financials; hopefully, you are using something better than payback or ROI. For ESCOs to survive and provide valuable services, they will probably have to change their traditional positive cash flow model based on just hard savings.

ADVANCEMENTS, ADVANCEMENTS, ADVANCEMENTS

There are so many developments, including adding the perception circle to the other five outer circles of the human centric lighting diagram, oral reading fluency (ORF) improvements of children, etc. Advancements will continue. So please attend my 'Human Centric Lighting' seminars through AEE and other organizations, read my upcoming white papers and magazine articles, and check updates on the Human Centric Lighting website (http://humancentriclighting.com/).

Appendix

Acronyms

AC	alternating current
AEE	Association of Energy Engineers
AMA	American Medical Association
ASHRAE	American Society of Heating, Refrigerating & Air Conditioning Engineers
BEF	ballast efficacy factor
BF	ballast factor
C	Celsius or Centigrade
CALIPER	commercially available LED product evaluation and reporting
CCT	correlated color temperature
CEE	Consortium of Energy Efficiency
CEM	Certified Energy Manager
CFL	compact fluorescent lamp
CLEP	Certified Lighting Efficiency Professional
CMH	ceramic metal halide
CQS	color quality scale
CRI	color rendering index
CW	cool white
DALI	digital addressable lighting interface
DC	direct current
DLC	DesignLights Consortium
DOE	Department of Energy
ECMH	electronically ballasted ceramic meta halide
EHID	electronically ballasted high intensity discharge
ESCO	energy services company
EVE	Equivalent Visual Efficiency
F	Fahrenheit
fc	foot-candle

GEB generic electronic ballast
HCL human centric lighting
HCLP Human Centric Lighting Professional
HID high intensity discharge
HPS high pressure sodium
IDA International Dark-Sky Association
IES Illuminating Engineering Society
ipRGC intrinsically photosensitive retinal ganglion cell
IS instant start
K Kelvin
KWH kilowatt Hours
LC lighting certification
LED light emitting diode
LEP light emitting plasma
LEED Leadership in Energy and Environmental Design
LPD lighting power density
LPW lumens per watt
lx lux
MH metal halide
NALMCO International Association of Lighting Management Companies
NCQLP National Council on Qualifications for the Lighting Professions
OLED organic light emitting diode
PAR parabolic aluminized reflector
PE Professional Engineer
PNNL Pacific Northwest National Laboratory
PS Program Start
QCS quality color scale
R reflector
RGB red, green, blue
RGB+ red, green, blue, and at least one more colored LED
ROI return on investment
S/P scotopic/photopic
SAD seasonal affective disorder
SSL solid state lighting

TM technical memorandum
 W watt
WW warm white

Glossary

1st Generation T8s (also called 700 series or basic grade)

Full wattage T8 lamps with CRI in the 70s and not the best lumens. F32T8s have about 2800 catalog or photopic lumens.

2nd Generation T8s

Mid-grade 800 series T8 lamps. Typically these F32T8s have 2950-3000 catalog or photopic lumens.

3rd Generation T8s (also called high performance or super)

Also called high performance, super, and high lumen 800 series T8 lamps. F32T8s that typically have 3100 catalog or photopic lumens.

3Q13

3rd quarter of 2013

4Q13

4th quarter of 2013

735, 841, 850, etc.

Nomenclature for many fluorescent lamps. First digit stands for CRI. 7 means CRI in the 70s. 8 means CRI in the 80s. Last two digits stand for Kelvin. 35 means 3500K; 41 means 4100K; 50 means 5000K.

800 Series

Fluorescent lamps with CRI in the 80s

A19

Typical household incandescent light bulb. A stands for arbitrary. 19/8″ (2 3/8″) is diameter.

Ambient lighting

General lighting, typically from ceiling mounted lighting fixtures

Amalgam

Material used in some fluorescent lamps to improve light output in cold and hot temperatures

Ballast

Required to operate fluorescent and HID lamps. There are both magnetic and electronic ballasts.

Beam spread (beam angle)

Angle of reflector lamps listed in degrees or terms, such as with spot and flood lamps

BER (ballast efficacy factor)

Way to compare how efficiently ballasts are driving the same number of same type of lamps. BEF = BF x 100/system wattage

BF (ballast factor)

How hard a ballast drives a lamp. Higher BF means more light and more wattage. Lower BF means less light and less wattage. Lamp lumen and wattage ratings in lamp catalogs based on 1.0 reference ballast.

Biax

Typically 2′ long, narrow U-bend fluorescent lamp

Candela

Light intensity in a direction. Mainly for reflector lamps.

CBCP (center beam candlepower)

Only for reflector lamps. Intensity (candela) at center of beam.

CCT (correlated color temperature)

Blackbody temperature in Kelvin degrees that looks like light source. (For example, as a piece of iron is heated it turns red, then yellow, white, and then bluish white.) Look at Kelvin.

CEM (Certified Energy Manager)

AEE's certification for qualified energy managers

CLEP (Certified Lighting Efficiency Professional)

AEE's certification for qualified lighting professionals

CMH (ceramic metal halide)

MH with very high CRI. Also called ceramic pulse start metal halide.

Cold cathode CFL

Type of CFL that has about double lamp life compared to typical CFLs.

Color quality scale

This may replace CRI, because it is better with LEDs.

Cones

One of the three photoreceptors in the human eye. Responsible for color discrimination. Photopic function.

CEE (Consortium of Energy Efficiency)

Lists high performance T8 lamps and ballasts, which many rebate programs use. See: www.cee1.org

CRI (color rendition index)

How natural that colors look. Daylight and incandescents are considered to have a perfect 100.

CW (cool white)
Most common CRI and Kelvin type for T12 lamps. 60 CRI & 4100K.

Cycles
With regard to rated lamp life. Typically 3 hours on, 20 minutes off, 3 hours on, etc. for fluorescents. Typically 10 hours on, 20 minutes off, 3 hours on, etc. for HID

Design lumens (or mean lumens)
Lumens at 40% of lamp life

Driver
Equivalent of a ballast for LEDs

Efficacy
This is fancy term for efficiency. Lumens per watt (like miles per gallon).

Energy saving F32T8s (or reduced wattage T8s)
These include 25, 28 and 30 watt lamps. There are several limitations with these lamps.

EOL (end of life)
This can be used for EOL lumens

F17T8
Fluorescent 17 watt Tubular 8/8" (1") diameter lamp (2' long)

F32T8
Fluorescent 32 watt Tubular 8/8" (1") diameter lamp (4' long)

F34T12
Fluorescent 34 watt Tubular 12/8" (1.5") diameter lamp (4' long). Formerly called energy saving F40T12 lamps.

F54T5HO

Fluorescent 54 watt Tubular 5/8" diameter High Output lamp (almost 4' long)

F96T12

Fluorescent 96" long Tubular 12/8" (1.5") diameter lamp. For some lamps, especially ones longer than 4', the number after F stand for length instead of wattage.

FC (foot-candles)

The amount of light that hits a target. What light meters measure. Horizontal fc are usually measured 30" above floor, which would be like putting light meter on a desk. Vertical fc are measured vertically like on a wall.

Fixture

Also known as lighting fixture or luminaire.

Fixture efficiency

Percentage of light that gets out of the fixture compared to light emitted by lamp(s).

Fluorescent

Lamp family in which the interior phosphor coating transforms UV energy into visible light.

Full spectrum lighting

Often a marketing-hype term used to sell lamps and bulbs at outrageous prices.

GEB (general electronic ballast)

Ballast that is not an extra efficient ballast.

Halogen

Incandescent lamp type with higher efficacy than common incandescents.

Halogen infrared
Higher efficacy than standard halogen.

Hibay (high bay)
Interior lighting fixtures often considered at least 25' high. Typical kind of lighting fixture in gyms and warehouses. Very similar to lowbay.

High performance
This typically relates to Consortium of Energy Efficiency's approved T8 lamps and ballasts.

HID (high intensity discharge)
Lamp category that includes MH (metal halide), HPS (high pressure sodium), and mercury vapor. Lamp catalogs often also include LPS (low pressure sodium).

HPS (high pressure sodium)
High intensity discharge lamp type that provides a yellow color, with about 22 CRI.

Illuminance
Density of light on a surface. Measured in foot-candles.

Induction
Fluorescent lamp type without cathodes, so really nothing to wear out. Typically 100,000-hour rated lamp life. Some lamp manufacturers include in HID section.

Incandescent
Lamp type in which the heated filament glows. Typical residential lamp type.

Initial Lumens
For lamps, the amount of lumens after 100 hours of operation

IS ballast (instant start)

Mainly for T8s. Ballast does not preheat lamp cathodes before starting lamp(s). Parallel wired so if one lamp burns out, remaining lamp(s) function normally. Most popular type of ballast for T8 lamps because of lower cost, higher efficiency, and fewer wires. Can shorten lamp life.

Kelvin

Temperature used to indicate correlated color temperature of a light source. Kelvin uses the same degree size as Centigrade. Zero for Kelvin is -273°C.

kW (kilowatt)

1000 watts.

kWh (kilowatt hour)

What the majority of utility electric bills are based on. Examples of 1 kWh are running a 1000W light bulb for 1 hour and running a 10W light bulb for 100 hours.

L70

Rated life of LED products when having 70% of initial lumens

Lamp life (rated life)

Rated lamp life is in the middle of a bell shape curve when half of the lamps have burned out and half are still working under laboratory conditions. See Cycles above for more information.

LC (lighting certified)

Certification by the NCQLP, the closest thing that the lighting industry has to PE (professional engineer) certification in the engineering realm. See: www.ncqlp.org

LED (light emitting diode)

One type of solid state lighting.

Lowbay

Interior lighting fixtures similar to hibay, but often considered less than 25' high. Typical kind of lighting fixture in gyms and warehouses.

Low voltage

Typically 12V

LM79

Established testing methodology to create a level field for product evaluation. It looks at 25°C ambient, power supply, stabilization, orientation, electrical instruments, and testing equipment. It defines what information is required—total light output, voltage, current, power, calculated efficacy, lumen distribution, CCT, CRI, spectral distribution, testing lab, and equipment used. LM-79 requires that solid state lighting products be tested to "absolute photometry." (Conventional HID/fluorescent uses "relative photometry.") Absolute photometry is lumen output of LED-based luminaires and is dependent on the chip, thermal management, drive current, and optical system. LED-based luminaires and lamps must be tested as a complete unit or system. Only DOE-recognized CALiPER testing laboratories results should be utilized.

LM80

Measure of lumen depreciation of LED light sources, arrays, and modules. Does not include complete products, nor define or provide complete product life.

LPS (low pressure sodium)

Ugly yellow light. Zero CRI. Common for exterior pole lighting in cities like San Jose and San Diego that have nearby observatories. Sometimes considered an HID.

Lumen

Amount of light that a light source generates in all directions. Is listed in lamp catalogs.

Lumen depreciation
Decrease in lumen output as lamp ages.

Luminance
Can be considered as brightness of an object or surface.

Luminaire
Fancy name for lighting fixture. Complete lighting unit consisting of housing, lamp sockets, lamp(s), ballast(s), and applicable reflector(s), louver(s), and lens(es).

Lux
About 10 lux equals 1 foot-candle.

MBCP (maximum beam candlepower)
Same as CBCP above.

Mean lumens (also design lumens)
Lumens at 40% of lamp life

Mercury Vapor
An HID lamp. Half as efficient as HPS and MH. Only good application is lighting pine trees.

MH (metal halide)
Good white light lamp in HID group. Includes standard/probe start, quartz PS start, & ceramic PS (CMH).

MR16
Multi Reflector 16/8" (2") diameter. 12V, which is low voltage.

NCQLP (National Council on Qualifications for the Lighting Professions)
See: www.ncqlp.org
Occupancy sensor

Control that automatically turns lights off based on no occupancy. Can also be set to automatically turn lights on.

PAR (parabolic aluminized reflector)
Typical for PAR38 and PAR30 halogen spots through floods.

Parabolic or parabolic louver
Often 4" x 4" openings that bare lamp can be seen through. Most common fixture type is 2' x 4' 18 parabolic cell troffer.

PCBs (polychlorinated biphenyls)
Very toxic substance in ballasts manufactured before 1980.

Photopic
Cone response in human eye.

PL
Common term for 5, 7, 9, or 13W two-pin, one-loop CFLs. PL is originally a Philips term. GE calls them BX. Sylvania calls them CF. Generic term is CFT(wattage)W/GX23/8XX.

Probe start MH
Same as standard MH.

PS ballast (program start or programmed start ballast)
Also called programmed rapid start. Improved version of rapid start. Preheats lamp cathodes before starting lamp(s). Usually series wired, so if one lamp burns out, other lamps also dim or turn off. More expensive and less efficient than IS ballasts.

Photometry (photometrics)
Measuring visible light or computer generated fc layouts.

PS MH (pulse start MH)
MH with starter in the lamp. Includes both quartz PS MH and ceramic PS MH (CMH).

Pupil lumens

Lumens based on scotopically or spectrally enhanced lighting.

Color quality scale

May replace CRI, because it includes bright or saturated colors.

Quartz

Incandescent lamp, the most common of which is 300W double ended, shaped like a pencil.

Quartz PS MH (quartz pulse start metal halide)

One of the two types of pulse start metal halide.

R (reflector)

The R is code for reflector lamps.

R9

Amount of red in LEDs

Restrike time

The time that a hot lamp requires to turn back on after it has been turned off. This can be 15 minutes for many MH lamps. Fluorescents have no restrike times.

Rods

One of the three photoreceptors in the human eye. Scotopic function.

RS ballast (rapid start)

Ballast that preheats lamp cathodes before starting lamp(s). Usually series wired, so if one lamp burns out, other lamps also dim or turn off. Being phased out and replaced with program start ballasts.

Scotopic

Rod response in human eye. Peak response in blue portion of the visible spectrum.

S/P (scotopic/photopic)

Ration available from lamp manufacturers. GE lists them in their paper and web catalogs.

SPD (spectral power distribution)

Graph of the radiant power emitted by a light source as a function of wavelength. SPDs provide a "finger print" of the color characteristics throughout the visible part of the spectrum.

Spectrally enhanced lighting

Formerly called scotopically enhanced lighting. Lighting with more blue content, which increases perceived brightness and acuity.

Standard MH (or probe start)

Older type of MH, also called probe start metal halide. Has the starter in the lamp.

Super T8s

Same as high performance, high lumen, and 3rd generation T8s.

Suspended indirect fixture

Suspended or pendant fixture that provides from significant to all uplight. This is very popular in new and remodeled office buildings

T5, T5HO, T8, T12, etc.

For these linear fluorescent lamps, T stands for tubular, and the number stands for how many eighths of an inch in diameter. HO stands for high output.

Task ambient lighting

Combination of task lighting and ambient lighting. With task lighting, the ambient lighting amount is reduced.

Task lighting

Lighting directed at desk or other specific surface/area to illuminate certain tasks like reading. Most common task light application is undercabinet in module office cubicles.

Task modified lumens

Lumens based on scotopically or spectrally enhanced lighting. Three classifications are brightness, paper tasks, and computer tasks.

Triple looper CFL

4-pin compact fluorescent lamp with three loops. Most common are 18, 26, 32, and 42W.

Troffer

Long, recessed lighting fixture with the bottom usually flush with ceiling. Most common type is 2' wide by 4' long, very typical in offices.

Tunable

Dimming and warm to cool white color changing. Also called dimming and Kelvin changing or shifting.

W (watt)

Unit of electrical power. Lamps and ballasts are rated in watts to show the rate at which they consume energy.

Warm-up time

The time it takes for a cold lamp to reach at least 90% of maximum light output, which can be quite long for some HID lamps.

Index